DIGITAL DESIGN:
From Gates to Intelligent Machines

LIMITED WARRANTY AND DISCLAIMER OF LIABILITY

THE CD-ROM THAT ACCOMPANIES THE BOOK MAY BE USED ON A SINGLE PC ONLY. THE LICENSE DOES NOT PERMIT THE USE ON A NETWORK (OF ANY KIND). YOU FURTHER AGREE THAT THIS LICENSE GRANTS PERMISSION TO USE THE PRODUCTS CONTAINED HEREIN, BUT DOES NOT GIVE YOU RIGHT OF OWNERSHIP TO ANY OF THE CONTENT OR PRODUCT CONTAINED ON THIS CD-ROM. USE OF THIRD-PARTY SOFTWARE CONTAINED ON THIS CD-ROM IS LIMITED TO AND SUBJECT TO LICENSING TERMS FOR THE RESPECTIVE PRODUCTS.

CHARLES RIVER MEDIA, INC. ("CRM") AND/OR ANYONE WHO HAS BEEN INVOLVED IN THE WRITING, CREATION, OR PRODUCTION OF THE ACCOMPANYING CODE ("THE SOFTWARE") OR THE THIRD-PARTY PRODUCTS CONTAINED ON THE CD-ROM OR TEXTUAL MATERIAL IN THE BOOK, CANNOT AND DO NOT WARRANT THE PERFORMANCE OR RESULTS THAT MAY BE OBTAINED BY USING THE SOFTWARE OR CONTENTS OF THE BOOK. THE AUTHOR AND PUBLISHER HAVE USED THEIR BEST EFFORTS TO ENSURE THE ACCURACY AND FUNCTIONALITY OF THE TEXTUAL MATERIAL AND PROGRAMS CONTAINED HEREIN. WE HOWEVER, MAKE NO WARRANTY OF ANY KIND, EXPRESS OR IMPLIED, REGARDING THE PERFORMANCE OF THESE PROGRAMS OR CONTENTS. THE SOFTWARE IS SOLD "AS IS" WITHOUT WARRANTY (EXCEPT FOR DEFECTIVE MATERIALS USED IN MANUFACTURING THE DISK OR DUE TO FAULTY WORKMANSHIP).

THE AUTHOR, THE PUBLISHER, DEVELOPERS OF THIRD-PARTY SOFTWARE, AND ANYONE INVOLVED IN THE PRODUCTION AND MANUFACTURING OF THIS WORK SHALL NOT BE LIABLE FOR DAMAGES OF ANY KIND ARISING OUT OF THE USE OF (OR THE INABILITY TO USE) THE PROGRAMS, SOURCE CODE, OR TEXTUAL MATERIAL CONTAINED IN THIS PUBLICATION. THIS INCLUDES, BUT IS NOT LIMITED TO, LOSS OF REVENUE OR PROFIT, OR OTHER INCIDENTAL OR CONSEQUENTIAL DAMAGES ARISING OUT OF THE USE OF THE PRODUCT.

THE SOLE REMEDY IN THE EVENT OF A CLAIM OF ANY KIND IS EXPRESSLY LIMITED TO REPLACEMENT OF THE BOOK AND/OR CD-ROM, AND ONLY AT THE DISCRETION OF CRM.

THE USE OF "IMPLIED WARRANTY" AND CERTAIN "EXCLUSIONS" VARIES FROM STATE TO STATE, AND MAY NOT APPLY TO THE PURCHASER OF THIS PRODUCT.

DIGITAL DESIGN:
From Gates to Intelligent Machines

BRUCE KATZ

DA VINCI ENGINEERING PRESS
Hingham, Massachusetts

Copyright 2006 by THOMSON DELMAR LEARNING. Published by DAVINCI ENGINEERING PRESS. All rights reserved.

Software contained on the CD-ROM was developed by Bruce Katz and is used with his permission.

No part of this publication may be reproduced in any way, stored in a retrieval system of any type, or transmitted by any means or media, electronic or mechanical, including, but not limited to, photocopy, recording, or scanning, without prior permission in writing from the publisher.

Cover Design: The Printed Image

DAVINCI ENGINEERING PRESS
CHARLES RIVER MEDIA, INC.
10 Downer Avenue
Hingham, Massachusetts 02043
781-740-0400
781-740-8816 (FAX)
info@charlesriver.com
www.charlesriver.com

This book is printed on acid-free paper.

Library of Congress Cataloging-in-Publication Data
Katz, Bruce F., 1959-
　Digital design from Gates to intelligent machines / Bruce F. Katz.—1st ed.
　　p. cm.
　Includes index.
　ISBN 1-58450-374-2 (alk. paper)
　1. Logic circuits. 2. Logic design. 3. Computer architecture. I. Title.
　TK7868.L6K384 2006
　621.39'5—dc22
　　　　　　　　　　　　2005022086

All brand names and product names mentioned in this book are trademarks or service marks of their respective companies. Any omission or misuse (of any kind) of service marks or trademarks should not be regarded as intent to infringe on the property of others. The publisher recognizes and respects all marks used by companies, manufacturers, and developers as a means to distinguish their products.

Printed in the United States of America
05 7 6 5 4 3 2 First Edition

CHARLES RIVER MEDIA titles are available for site license or bulk purchase by institutions, user groups, corporations, etc. For additional information, please contact the Special Sales Department at 781-740-0400.

Requests for replacement of a defective CD-ROM must be accompanied by the original disc, your mailing address, telephone number, date of purchase and purchase price. Please state the nature of the problem, and send the information to CHARLES RIVER MEDIA, INC., 10 Downer Avenue, Hingham, Massachusetts 02043. CRM's sole obligation to the purchaser is to replace the disc, based on defective materials or faulty workmanship, but not on the operation or functionality of the product.

Contents

Preface	xvii

1 Numbers and Number Systems — 1

Introduction	1
Numbers	4
Positional Number Systems and Bases	5
Conversions Between Bases	8
Conversions to and from Other Bases to Base 10	9
Binary Number Systems	11
Binary Addition and Subtraction	12
Binary Multiplication	15
Negative Numbers in Binary: Signed Magnitude and Two's Complement	15
Codes	18
BCD	18
Gray Coding	19
Parity	20
ASCII and Unicode	20
Summary	22
Exercises	22
LATTICE Exercises	23

2 Boolean Algebra — 25

Introduction	25
Logical Functions in Boolean Algebra	26
Truth Tables	29

        Tautology, Equivalence, and Logical Laws 31
        Other Useful Logical Operators 33
    Simplification 34
        Simplification with Logical Laws 35
        Minterms and Maxterms 37
        Karnaugh Maps and Minimization 42
    Summary 50
    Exercises 51
        LATTICE Exercises 53

3 **Elementary Combinational Circuits** 55
    Introduction 55
    Logic, Gates, and Circuits 56
        Elementary Gates 56
        Circuits to Functions and Truth Tables 59
        Realizing a Function Directly 61
        Realizing a Circuit Through Minterms and Maxterms 63
        Alternative Representations of SOP and POS Functions 64
        Realizing a Minimized Form of a Function 66
    Gates and Integrated Circuits in Practice 67
        Logic Technologies and Logic Families 68
        Values and Voltages 69
        Fan-In and Fan-Out 70
        Gate Delays and Circuit Delays 72
        Implementation of Gates 74
    Summary 78
    Exercises 78
        LATTICE Exercises 80

4 **Complex Combinational Circuits** 81
    Introduction 81
    Binary Adders 82

Full Adder	82
Ripple-Carry Adder	85
Carry–Look-Ahead Adder	86
Two's Complement Addition and Subtraction	88
Decoders and Encoders	90
Binary Decoders	91
Decoder Applications	95
Encoders	97
Multiplexers and Demultiplexers	100
Programmable Logic Devices (PLDs)	107
Programmable Read Only Memory (PROM)	107
Programmable Array Logic (PAL®)	108
Programmable Logic Array (PLA)	109
Summary	112
Exercises	114
LATTICE Exercises	116

5 Elements of Sequential Design — 117

Introduction	117
Latches	119
SR Latch	119
D Latch	123
Flip-Flops	125
Edge-Triggered D Flip-Flop	125
Edge-Triggered J-K Flip-Flop	127
Registers	130
Parallel-Load Registers	130
Shift Registers	132
Summary	134
Exercises	134
LATTICE Exercises	137

6 Sequential Machines — 139

- Introduction — 139
- Finite State Machines — 140
- Mealy and Moore Machines — 144
- Sequential Machine Analysis — 145
- Sequential Machine Synthesis — 150
  - General Method — 150
  - The Parity Example — 153
  - A Sequence Recognition Example — 157
  - A Maze Example — 162
- Designing with J-K Flip-Flops — 170
- Summary — 174
- Exercises — 175
  - LATTICE Exercises — 177

7 Elements of Computer Design — 179

- Introduction — 179
- Computer Organization — 182
- Memory — 183
- The CPU — 189
- I/O — 191
- Summary — 193
- Exercises — 194
  - LATTICE Exercises — 195

8 The Design of a Simple CPU and Computer — 197

- Introduction — 197
- The Register Set — 198
- The Instruction Set — 201
- The Control Unit — 204
  - The Fetch-Decode-Execute Cycle — 205
  - The Control Unit Finite State Machine — 207

Data Paths	210
The ALU	214
Putting It All Together	218
Further Issues in Computer Design	223
Microsequencing	224
Interrupts	224
RISC and Pipelining	225
High-Level Languages	226
Summary	226
Exercises	227
LATTICE Exercises	228

9 Explorations in Digital Intellligence — 229

Introduction	229
Pattern Recognition	232
Pattern Completion	235
Interference and Expert Systems	236
Neural Networks	240
Learning	249
Search	254
Emergent Behavior	256
Summary	259
LATTICE Exercises	260

Appendix   The LATTICE System — 263

Introduction	263
Installation	264
System Requirements	264
Installation Procedure	264
General Operation	265
Program Layout	265

File Menu Options	265
Animation Options	266
Mouse Buttons	266
Truth Systems	267
Variable Drop-Down Box	267
Operator Drop-Down Box	268
Truth Table Options (Bottom of Screen)	269
Variable Settings Dialog Box	269
State Systems	270
Column 1 (State Color)	271
Column 2 (State Action)	271
Column 3 (State Name)	273
Columns 4 Through the End of the Table (State Transitions)	273
State Variable Settings	273
State Table Options (Bottom of Screen)	275
Tricks of the Trade	275
System Submission	277
Index	**279**

Preface

For a variety of reasons, but most significantly, the rise of the computer as a central, if not dominating, force in our lives, digital circuits have taken on a new importance in the past twenty years. There are abundant and various resources on the software that drives these machines, and whole sections of bookstores are devoted to the topic. But for the student who wishes to understand what really makes computers and their simpler digital counterparts work, the choices are limited to a set of well-meaning, but dry, textbooks on digital design.

This book aims to remedy this problem in two ways. First, it explicitly acknowledges that digital design is an intrinsically visual and dynamic process. It is visual in the sense that the end products (digital circuits), are physical objects best represented in diagrammatic form. Above all, it is visual in the sense that for the two intellectual underpinnings of the design process (truth tables and finite state diagrams), the most direct and instinctual representations are two-dimensional drawings.

Therefore, most books written on the topic are liberally populated with figures. This captures the visual aspect of the digital design, but does little to address the dynamic aspect. To take just one example, a truth table is both a visual *and* dynamic object. It is trivially visual in the sense that it is a table, but it is also dynamic in the sense that it ought to be possible to quickly change the connectives or the variables or the expressions themselves to see the effect on the truth values. This, of course, is not possible on the written page, but is eminently suited to a program running on a digital machine. It is ironic, but not without precedent, that the teaching methods for a given topic lag behind the technology that the teaching is supposed to be about.

This book closes the gap between pedagogy and technology with the inclusion of the LATTICE software system. LATTICE contains truth tables and state tables for simulating combinational and sequential circuits, respectively. These tables drive cellular automata to provide a direct visual counterpart to the dynamics that they represent. This differs significantly from other simulation techniques in that (1) it is easier to use—the student can be up and running with minimal effort, and (2) it abstracts the notion of the underlying logic and dynamics from the particular implementation of such. This is not to say that circuit realization is not important, and in fact, this process is at the heart of this book. Rather, it provides an entry to what is a daunting topic for many by providing a direct animated counterpart to the topic at hand. It is also hoped that the emergent visual and musical behavior of cellular automata offers a relatively painless environment whereby difficult material can be absorbed unconsciously and with relatively little effort.

The other main difference between this book and competing texts is that the current effort aims to be comprehensive without presenting an overwhelming amount of material. An introductory textbook in the field of digital design can aim at two distinct and competing goals. The first is to provide the student with the foundation for future progress in the field, and the second is to serve as a comprehensive reference for future use. It is not hard to imagine a professor putting the finishing touches on his 900-page tome all the while dreaming of a working engineer some twenty years after graduation saying to himself, "Hmm, tricky problem. I think I'll consult my well-worn Cholmondesley from freshman design on this one." However gratifying this imaginative diversion may be, my experience has been the beginning engineer will suffer by this approach, even if he is later compensated. The reason is that there is simply too much material to grasp in a single course. It would be as if a Physics 101 book provided a detailed account of Newtonian mechanics *and* quantum mechanics. Normally, the former is provided in detail and the latter, if at all, in outline only.

The mind of the budding engineer is not different from the rising physicist in the sense that it needs to understand the nature of the problem before it can grasp the intricacies of the solution. What is (are) the problem(s) of digital design? They are twofold. The first is to understand how a device that implements a logical function can be built, and the second is the design and implementation of a sequential machine. Neither of these is difficult for the student with the proper background, but it would be a mistake to assume that this background is firmly in place for all students. Logic is the foundation for the former and it is poorly taught, if at all, in the secondary school system, and is not always given a formal introduction in undergraduate education. Discrete mathematics in general, and finite state machines in particular, are the foundation for the latter, and this also too often falls victim to spotty treatment. In both cases, the LATTICE system provides a means of making these abstract endeavors direct and concrete.

Finally, and less obvious, the problem of digital design can no longer be held separate from the design of intelligent machines. As argued in the final chapter, so-called smart devices are increasingly common. It is not possible however, to begin to design such devices without a grasp of the foundations of intelligent machines. This topic is invariably skipped in an introductory text because it's construed as either too advanced or too esoteric. Yet as the final chapter shows, many of the concepts are extensions of what has already been developed in earlier chapters. One cannot ignore the motivational properties of such topics. It is one thing to build a mod 5 counter, and quite another to build a machine that uses the principles of genetic algorithms to design a better musical composition. The excitement that the student feels cannot be held separate from the educational process, and it is hoped that this excitement will facilitate the grasping of the primary material, and ultimately provide the desire to create the next generation of digital machines.

ACKNOWLEDGMENTS

I would like to thank the many people I have taught digital design with over the years including S. "Basu" Basavaiah, Tim Kurzweg, Alex Meystel, Al Tonik, Oleh Tretiak, and Lazar Trachtenburg. Other colleagues, especially Allon Guez, also contributed valuable suggestions with respect to the accompanying software. They provided valuable insights into the subject and often influenced this book without realizing it. I would also like to thank the many hundreds of students who have contributed by suffering through my pedagogical experiments until I arrived at an adequate way of presenting this material, and especially Dan Lofaro for his help with the music files. Finally, I would like to thank my wife, Ayça who suffered, by putting up with me as I struggled to complete this work and by listening to the unusual maqams generated by the LATTICE program.

1 Numbers and Number Systems

In This Chapter

- Introduction
- Numbers
- Positional Number Systems and Bases
- Binary Number Systems
- Codes
- Summary
- Exercises

INTRODUCTION

Digital design is the study of circuits that work with quantized signals. Quantized means that the signals are channeled into two discrete levels, represented by 0 and 1. The alternative is an analog signal, or one that can take a continuous range of values. For example, sounds are analog signals: at any time, the sound pressure, represented in decibels, can be at an arbitrary level. If the sound is recorded onto a medium such as tape, then the analog nature of the sound will be preserved. However, if it is stored on a CD-ROM it will be converted into a stream of 1s and 0s first and some information may be lost in the process. This is why some musical purists prefer media such as tapes and vinyl records, despite the difficulties with storage and maintenance.

Why then, do we use digital signals? One reason is resistance to noise. For example, consider adding a noise level of 0.073 to an analog signal. The signal will be distorted by exactly that amount. However, if this quantity is added to a digital signal of 0 or 1, and we assume that the system produces the nearest digit (0 or 1), then the noise has no effect. In the first case, 0 goes to 0.073, which is still much closer to 0 than 1, and in the second case, 1 goes to 1.073, which is much closer to 1 than 0. Thus, a small amount of noise has no effect on the signal.

However, the most important reason that digital signals and digital circuits have largely eclipsed their analog counterparts, is the ease with which computations can be performed on such signals. While there are still some specialized analog devices to perform purely mathematical computations quickly, the vast majority of computational machines are digital. Furthermore, digital devices are rapidly overtaking analog ones in domains where the latter was formerly dominant, such as music, radio, television, and wireless devices (such as cell phones). The primary reason for this is versatility. In principle, a digital device is capable of performing arbitrary computations. This includes both logical and mathematical operations.

However, it is not at all obvious that digital computation is a natural fit for logic, and assuming that it is, that logical computations can be transformed into mathematical ones. One of the many goals of this book is to show that both of these propositions are the case. Chapters 1–7 lay the foundation for these conclusions. By the time we reach Chapter 8, we will be able to construct an elementary computer that performs both logical and mathematical operations on the contents of its memory. The final chapter not only shows that routine calculations are possible within the digital paradigm, but also that many of the elements of intelligence fall within the province of the digital domain.

Of the many ways to enhance the subject of digital design with digital means, this book takes a somewhat different approach—one that we think you will find challenging and stimulating. Digital gates and circuits are illustrated with logical automata. The program that embodies these entities is know as LATTICE (Logical AuTomaTa Integrated Creation Environment). If you have not already done so, install the LATTICE program on your machine by clicking on the install.exe icon on the accompanying CD-ROM, and follow the directions. The Appendix describes the installation process in more detail and contains a manual for the program. Logical automata are extensions of cellular automata, simple machines that are characterized by two states, referred to as alive and dead, and which compute their next state as a function of the states of the automata in their immediate vicinity. The "logical" aspect of the LATTICE system derives from the fact that the automaton response is a logical function of its surroundings rather than simply a function of the count of the number of living neighbors, as in more typical cellular automata systems.

At this point, it is not necessary to understand all the workings of LATTICE, but it will be helpful to experiment with a sample application. Click on the LATTICE shortcut on your desktop or open the program from the Start menu. Use the pull-down menus at the top of the application window to open the "life" system (File>Open system...) in the Chapter1 folder (in the LATTICE examples folder in C:\Documents and Settings\All Users\Application Data\Lattice). This system reproduces the *Game of Life*, invented by the Princeton mathematician John Conway in 1970. The rules of the game are illustrated in Figure 1.1, which shows two examples for each rule application. For the purposes of this illustration, a dark cell is alive and a white one is dead. Rule 1 states that if exactly three of the eight cells surrounding a given cell (the marked center cell) are alive, then the marked cell comes alive in the next generation. Rule 2 states that if a cell is already alive and either two or three of its eight neighbors are alive, then the cell will continue to live in the next generation. In all other cases the center cell dies. In each generation, these rules are applied to every cell in the lattice on the basis of the state of the system in the previous generation.

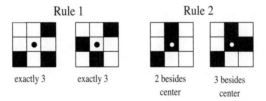

FIGURE 1.1 Examples of rule applications for the *Game of Life*. Rule 1: exactly three live cells in the eight-neighborhood surround (e.g., the three dark cells) generate life. Rule 2: Two or three in the eight-neighborhood surround allow life to continue for the existing living center cell.

Click on the double arrow icon below the lattice to start the simulation. You may stop the simulation at any point by clicking on the leftmost icon; it is also possible to change the state of the cells while the simulation is running by clicking on a cell or by dragging over multiple cells. You can also wait until this system stops changing (after about 1,200 cycles) and "restart" the animation by clicking near the stable alive cells. Notice that a single click can often alter the behavior of the system radically. This aspect of the system, its sensitivity to initial conditions, as well as the fact that complex patterns emerge from a system with exceedingly simple rules, will be discussed in more detail in Chapter 9. There is also an abundant amount of information regarding the *Game of Life* on the Web; a good place to start is *http://www.math.com/students/wonders/life/life.html*.

In the *Game of Life*, a cell is either alive or dead. LATTICE also uses cells that can be in more than two states as a means of illustrating sequential machines (see Chapter 6). Additionally, these cells optionally generate another output signal when a given state is achieved. Load the "music" example from the Chapter1 folder for an example of the system that generates sounds as different states are reached. This is a system with four machines: a background, a single cell that keeps the beat, a single cell for the bass, and a single cell for other notes. Depending on the background that surrounds these cells, they will transition to different states. Many of the states are linked to sound files, which are played as the state is reached. Click on the double forward arrow to see the machine at work, and advance the speed slider to the maximum position (make sure your speakers are on). You may also click on the cells surrounding the single cells to alter the state transitions and therefore alter the sound sequences. The periodicity inherent in the system leads to realistic-sounding music, and this periodicity is an emergent property of the system; that is, one that is a function of all the parts working together, but not necessarily present in the individual partitions. We will discuss this more fully in the closing section of Chapter 8.

NUMBERS

Before approaching digital design proper, we need to have a thorough understanding of two foundational areas. The first is number systems, which is the topic of this chapter, and the second is elementary logic, which forms the subject matter of Chapter 2.

Numbers are objects that we usually do not think about, probably because most of us have seen them since a young age and now take them for granted. Yet, as this chapter will show, there is a great deal of complexity lurking behind this seemingly innocuous topic, the mastery of which is essential to the design of digital circuits and computer systems. Let us begin by considering numerals (symbols that represent quantities) and number systems (groups of numerals). The simplest such system would be to have a mark for each element. For example, "|||" would represent the quantity we would ordinarily represent with the Arabic numeral 3, and "|||||" would represent 5. There is nothing intrinsically wrong with this representation, but like all representations, it is only as good as the job or jobs it is intended for. In this case, the system breaks down once the quantity exceeds a dozen or so. Consider trying to represent the quantity 1,024, a number that comes up often because it is equivalent to 2^{10}. The four successive digits, 1, 0, 2, and 4, which we would ordinarily use, would need to be replaced with 1,024 marks. It is easy to see how this system would get out of hand quickly.

An alternative system with the virtue of compactness is the Roman numeral system. In this system, successive symbols represent growing quantities, allowing the representation of large numbers with relatively few numerals. For example, the quantity we would ordinarily represent as 1,973 can be represented by MCMLXXIII, something that you might see at the end of a movie indicating the year it was produced. As you can see, the Roman system is not quite as compact as our ordinary representation, but that is not the problem. The real problem occurs when one is attempting to perform an arithmetic operation with this system. Consider adding 1,024 (MXXIV) to 1,973 (MCMLXXIII). One cannot add in columns as we would ordinarily do; nor can one simply append the two numbers because some Roman numerals subtract rather than add to the quantity, as we recall from grade school. Other operations such as multiplication are even more problematic.

What is clearly needed is a representation that is both compact and amenable to the typical mathematical operations that one would need to perform on these representations. As we will see, there is no ideal system for all operations. Nevertheless, ordinary positional number systems meet many of the most common needs. Therefore, we will consider them first.

POSITIONAL NUMBER SYSTEMS AND BASES

Consider again the number 1,024. The quantity that this represents is actually formed by an implicit computation: 1*1000 + 0*100 + 2*10 + 4*1. Rather than having a unique symbol for different quantities, positional number systems such as this rely on the position of a fixed number of symbols to indicate magnitude. For example, the 1 at the front of the number represents not a single item but a quantity of 1000 because it is the fourth and leftmost numeral. Furthermore, it is trivial to add 1,024 to 1,973 using the methods that we learned in grade school, that is, by adding up the individual columns and carrying an intermediate result to the next column when necessary. In summary, positional number systems achieve compactness by virtue of multiplying a small collection of numerals by a set of fixed quantities that grow as a function of position, and they achieve ease of computation by virtue of repeating a relatively simple operation on one column at a time. Therefore, they are ideal for representation and computation by both humans and machines.

The numbers that we are most used to working with are in the decimal system. However, the decimal system, or the base 10 system, is only one of many positional systems that we can use. Computers for example, compute primarily in base 2, or the binary system. In general, a number in a positional system with base b (or radix), will represent the following quantity:

$$a_n b^n + a_{n-1} b^{n-1} + a_{n-2} b^{n-2} + ... + a_0 b^0 + a_{-1} b^{-1} + ... + a_{-m+1} b^{-m+1} + a_{-m} b^{-m} \quad (1.1)$$

This expression includes both integral components (the powers of the base b above 0) and fractional components (the powers of the base b below 0). Each digit a_i in the number represents a weight, which is multiplied by the base b raised to an exponent, which depends on the position of the digit.

This is best illustrated by an example. Suppose the number is 417.23_8. The subscript 8 indicates that this is base 8; if there is no subscript, then the base is the default base of 10. This number is the quantity $4*8^2 + 1*8^1 + 7*8^0 + 2*8^{-1} + 3*8^{-2}$, or $256 + 8 + 7 + 2/8 + 3/64$, or 271.296875 in decimal. Thus, numbers in any base are constructed similarly to those in the familiar base of 10; the only thing that changes is the base of the exponent of each component.

Please note the following additional characteristics of any number regardless of base:

- The point in a number with a fractional component is called the radix point in general, or the Greek-language equivalent of the base (for example, in the base 8 number above, the point is called an octal point).
- The digits to the left of the radix point represent the weights of the integral (nonfractional) component of the number. Of special interest is the digit immediately to the left of the radix point. Because any number raised to the 0 power will be 1, this digit always represents the 1s place, regardless of the base.
- The digits to the right of the radix point represent fractional quantities. These correspond to the negative powers in Equation 1.1. Recall that a number raised to a negative power is simply the inverse of the number raised to the same positive power.

Numbers and Number Systems

The most commonly used nondecimal bases are shown in Table 1.1, which shows the quantities 0 through 16 in decimal and these bases. The table illustrates a number of features of counting in nondecimal bases. First, the representation for a base b number will be identical to the representation in base 10 up to the index of the base. For example, the numbers 0 through 7 are the same in base 8 and base 10. However, when we reach 8, the octal number becomes 10. Why? Because 10_8 is $1*8^1 + 0*8^0$, which equals 8. The same is true for binary, although in this case extra digits are necessary much sooner because we only have 1 and 0 to play with. By the time 16 is reached, the binary equivalent requires five digits. The other feature that the table illustrates is that bases greater than 10 require extra symbols. For example, the quantities 10 through 15 in hexadecimal are by convention represented by the symbols A through F respectively. Other than this, hexadecimal works the same way as any other base.

TABLE 1.1 Commonly Used Bases and Their Decimal Equivalents

Decimal (base 10)	binary (base 2)	octal (base 8)	hexadecimal (base 16)
0	00000	0	0
1	00001	1	1
2	00010	2	2
3	00011	3	3
4	00100	4	4
5	00101	5	5
6	00110	6	6
7	00111	7	7
8	01000	10	8
9	01001	11	9
10	01010	12	A
11	01011	13	B
12	01100	14	C
13	01101	15	D
14	01110	16	E
15	01111	17	F
16	10000	20	10

Which base is best for people to use? In most cultures, the counting system was most likely inspired by the fact that we have ten fingers and toes, and thus base 10 is the most commonly used base. However, from a mathematical standpoint, base 12 is probably better because 12 has four divisors other than itself and 1 (2, 3, 4, and 6), and 10 has only 2 (2 and 5). The more factors, the easier it is to represent fractional quantities. For example, if you want 1/3 dozen doughnuts, you get an even four doughnuts. Thus, it is not an accident that base 12 spontaneously appears in spite of the prevalent decimal system: there are 12 hours on the clock, 12 inches to the yard, 12 pence to the shilling in the pre-decimalized British currency system, and of course, 12 months in a year.

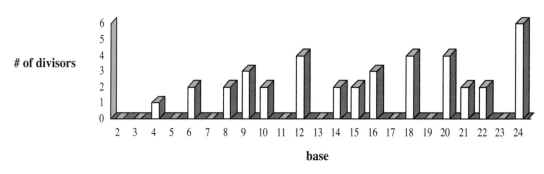

FIGURE 1.2 The number of divisors for each base up to 24.

Other possible bases are indicated by the graph in Figure 1.2, which shows the number of divisors for bases up to 24. Both 18 and 20 have four divisors, and 24 has 6. The problem with bases much higher than 12, however, is that they require an increasing number of symbols, and this makes them more difficult to learn and manipulate, especially for children (base 16 is common in the computational world, as we will see, because is useful for representing bytes and sets of bytes). If it were not for this fact, we might as well use base 60 (divisible by 2, 3, 4, 5, 6, 10, 12, 15, 20, and 30). Base 12 needs, in contrast, only two extra symbols relative to base 10. It is unlikely, however, that base 12 will ever replace base 10, although there are some diehard proponents of this system (see the Dozenal Society's Web site at *http://www.polar.sunynassau.edu/~dozenal* for more info).

Conversions Between Bases

If one base b1 happens to be an integer power of another base b2, then there is a relatively easy way to convert between numbers in these bases. This can be derived using the fact that an n-digit number in base b1 is capable of representing b1^n quantities, if zero is included. For example, in base 2, a 3-digit number can represent $2^3 = 8$ quantities, the numbers from 0 to 7 (see Table 1.1). The same quantities can be represented by a single octal number, as can also be seen from the table (Why? $8^1 = 8$). The question we need to answer, in general, is how many digits it takes to represent the same set of numbers in base b2 as in base b1 given that one is an integer power of another. That is, we need to solve for m in

$$b1^n = b2^m. \quad (1.2)$$

Here b1 is the smaller base, n is the number of digits in this base, and m is the number of digits in base b2. But we know, by hypothesis, that b2 is an integer power of b1, say

$$b2 = b1^p. \quad (1.3)$$

Substituting Equation 1.3 into Equation 1.2, we get

$$b1^n = (b1^p)^m = b1^{pm}, \quad (1.4)$$

which implies that $n = pm$, or that $m = n/p$. This means that for every n digits in the smaller base, b1, we need n/p digits in the second base. Or equivalently, that grouping p digits together in the first number will produce each successive digit in the second number.

This conversion process is perhaps easier to understand by example. Let's say we have 101111_2 and we wish to convert this to base 8. Because $2^3 = 8$, we can group the binary digits (bits) by threes to produce $\{101\}\{111\}_2$. However, 101 is just 5 in base 8, and 111 is 7, so $101111_2 = 57_8$. Alternatively, suppose we need to convert 101110_2 to base 16. We need to group by fours ($2^4 = 16$) in this case, although this time we have only six digits. The trick is to add two leading 0s to yield 00101110_2 (leading 0s never change the value of a quantity), and then the grouping yields $\{0010\}\{1110\}_2$, or $2E_{16}$, as can be readily verified from Table 1.1. Let us examine one more case in the opposite direction. Suppose that the number is $F4_{16}$. What is the equivalent in binary? Here the trick is to decompose each hex digit into its binary equivalent, which will contain four digits. Thus, $F4_{16}$ is $\{1111\}\{0100\}_2$, or 11110100_2 after removing the braces.

Conversions to and from Other Bases to Base 10

It is not possible to use the previous tricks to convert to decimal because 10 is not usually an integer power of any number (the trick could be used in the rare instance one might want to convert from decimal to base 100, though). However, conversion from any base to decimal is trivial because if it follows directly from the positional character of the number system as expressed by Equation 1.1. For example, suppose you wanted to know what 11101.101_2 was in decimal. By Equation 1.1 this is simply $1*2^4 + 1*2^3 + 1*2^2 + 0*2^1 + 1*2^0 + 1*2^{-1} + 0*2^{-2} + 1*2^{-3} = 29 + 0.5 + 0.125 = 29.625$.

Conversion from decimal to another base is a bit more complicated. One method, which might be termed the "engineer's" approach because it follows from the intuitive grasp of positional number systems that comes with experience, is to simply determine the highest quantity in the target base that is less than the given number, subtract this from the number, and repeat the process. For example, suppose we want to express 67 in binary. In this base, we can only work with powers of 2, and the largest power of 2 that is still less than 67 is 64, or 2^6. Thus, we need to put a 1 in the seventh position (recall that we start counting from 0), and the binary number will look something like $1XXXXXX_2$. Now we need to represent the difference between 67 and 64, or 3. The highest power of 2 less than 3 is 2^1, and thus the number at this point will be $100001X_2$. Finally, we need to represent $3 - 2 = 1$, which just means a 1 in the first position, yielding a final representation of 1000011_2.

A more formal method follows from the rerepresentation of the integral part of a number in base b from Equation 1.1 to

$$[(((a_n b + a_{n-1})b + a_{n-2})b + ...a_1]b + a_0. \qquad (1.5)$$

In this expression, each coefficient a_i picks up i b's to yield a final term in the sum $a_i b^i$, as in the original formulation. The point of this representation is that successive divisions by b will yield the a's as remainders, in the order from least significant digit to most (which is what we need to represent the number in that base delete this phrase). We can see this by first dividing Equation 1.5 by b; this yields a remainder of a_0. If we then take the dividend of that operation (everything in the brackets in Equation 1.5), and divide by b, we get a_1 as a remainder, and it is easy to see that each successive division will yield the next a_i.

Again, this process is best illustrated by example. Suppose we need to convert 125 to base 3. We accomplish this as follows:

$$125/3 = 41 \quad \text{remainder} \quad 2$$
$$41/3 = 13 \quad \text{remainder} \quad 2$$
$$13/3 = 4 \quad \text{remainder} \quad 1$$
$$4/3 = 1 \quad \text{remainder} \quad 1$$
$$1/3 = 0 \quad \text{remainder} \quad 1$$

The answer is just the remainder column in reverse order, or 11122_3 (check that this is correct by converting back into decimal). There is a similar procedure for converting the fractional part of a decimal number into another base, although we will not consider it here.

BINARY NUMBER SYSTEMS

Of all the nondecimal systems, the base 2 or binary system is the most important with regard to digital computation. This is because of the nice correspondence between base 2 and logic, as discussed in Chapter 2, and also because of the relative ease of constructing circuits that operate on binary numbers, as discussed in Chapter 3. Because of its ubiquity, it is necessary to become familiar with some common properties of base 2 and how arithmetic operations are carried out in this base.

One aspect of binary that all students of digital design should have firmly tucked away in their mental recesses, are the most commonly occurring powers of two, as listed in Table 1.2. Eight bits ordinarily constitutes a byte and can represent the quantities between 0 and 255, or alternatively, a single ASCII character as discussed below. Four bits is called a "nibble," and 16 binary digits constitute one ordinary UNICODE character, also discussed in more detail below. Two to the tenth power (2^{10}) is conveniently close to 1,000; therefore, the quantity 1,024 is abbreviated 1 K. For example, 1 K bytes are 1,024 bytes, or a total of 8,184 bits. The quantities 2^{20}, 2^{30}, and 2^{40} are abbreviated 1 M (mega), 1 G (giga), and 1 T (tera), respectively. An easy way of computing powers of two is to use this table and the fact that multiplying numbers with the same base entails adding exponents. For example, 2^{24} is $2^{20}*2^4 = 1,048,576 * 16 = 16,777,216$.

TABLE 1.2 Commonly Occurring Powers of Two and Their Significance

n	2^n	Significance
4	16	one nibble (1/2 a byte)
8	256	one byte; one ASCII char
10	1,024	quantity abbreviated 1 K (kilo)
16	65,536	two bytes; one UNICODE char
20	1,048,576	quantity abbreviated 1 M (mega)
30	1,073,741,824	quantity abbreviated 1 G (giga)
40	1.099E12	quantity abbreviated 1 T (tera)

A couple of words of caution are in order with respect to Table 1.2. First, notice that 1 K is not quite the same as 1,000. This means that as the powers of two grow, they are increasingly larger than the corresponding even power of 10. For example, if you are lucky enough to have a hard drive that holds 1 terabyte of data, it actually holds about 10% more than 1 trillion bytes. Second, these abbreviations apply primarily to storage. When talking about speed, the even powers of 10 apply. Thus, if your computer chip is running at 1 GHz, then it processes an instruction exactly 1 billion times a second, not 1.073 billion times a second.

There are two additional items that should be thoroughly mastered before moving on to more complex matters. The first is the notion of a full count in binary. This occurs when all the bits are 1. For example, a nibble at full count is 1,111. A full count will always correspond to the quantity $2^n - 1$, where n is the number of bits present (check that the nibble corresponds to 15, or $2^4 - 1$). From this, it is easy to see that a total of n bits is capable of representing 2^n quantities. The quantities from 1 to the full count will be $2^n - 1$, and if we include one more for 0, we get $(2^n - 1) + 1$ or just 2^n (please note that we always start counting from 0 in digital design, and also in computer science as a whole). Taking the nibble as an example again, and referring to Table 1.1, we see that 1 through 1,111 represents 15 quantities, and including 0 makes this a total of 16 or 2^4.

Table 1.1 also reveals an additional regularity of interest—every set of 2^n numbers are repeated, first with a preceding 0 and then with a preceding 1. For example, look at the first $2^2 = 4$ numbers, starting from 0 as usual. The sequence is 00, 01, 11, 00. Now look at the next four numbers representing the decimal quantities 4 through 7. They are identical in the first two places, and differ only by the addition of a 1 in the most significant (leftmost nonzero) place. This symmetry will prove useful in the construction of some digital circuits, as described in later chapters.

Binary Addition and Subtraction

In principle, binary addition and subtraction is no different from the decimal operations that we learned in second grade. In both cases, it is a column by column operation, and the only significant points of interest are carries in the addition and borrows in the subtraction. An illustrative example of the former appears below:

```
  00110010      carry
  10011101      addend1
+ 01011001      addend2
  --------
  11110110      sum
```

The sum is between the two addends; the carry row is a bookkeeping row for the purposes of the calculation only, although 1s in this row are added to the total just as if they appeared in the addends. The least significant bit (rightmost) in this row is always 0. Then, addition proceeds column by column, with $1_2 + 1_2 = 10_2$, that is, a 0 is put in the sum row and the carry is 1. If the entire column is 1s, then the sum

Numbers and Number Systems   **13**

is $1_2 + 1_2 + 1_2 = 11_2$, that is, a 1 is put in the sum row and a 1 is also put in the carry column. As with all binary operations, you can always check that you produced the right result by converting everything to decimal (in this case, $157 + 89 = 246$).

ON THE CD

As a foreshadowing of things to come, examine the addition example from Chapter 1 on the CD-ROM. This example reproduces the addition example above, that is, $10011101 + 01011001$. A 1 is represented by a live or shaded cell and a 0 is represented by a clear cell. To perform the addition, click on the step icon (the single right arrow) until the correct result (11110110) appears below the addends. This is a general system that works for all combinations of addends; try entering some other numbers to check this (the number of steps until the system reaches a steady state, thereby indicating it has finished the computation, will depend on the numbers that you have entered). We have not yet developed enough theory to explain how this addition machine works, but it is worth making an important remark at this point. LATTICE contains no arithmetic operations per se, only logical ones. It performs addition by reducing arithmetic operations to logical operations. Likewise, digital circuits perform arithmetic by applying the appropriate logical operators to binary representations of numeric quantities. We will cover this in more detail in Chapter 4.

An illustrative example of the binary subtraction between two bytes appears below.

```
  01000110      borrow
  11011100      minuend
+ 01101001      subtrahend
  --------
  01110011      difference
```

As in the case of addition, the borrow row is indicated for bookkeeping purposes, and the actual subtraction is between the minuend and the subtrahend. Borrows occur whenever the net difference would be negative. For example, in the rightmost column, the operation is $0_2 - 1_2$. Thus, a borrow is needed from the second column, effectively making the operation $10_2 - 1_2$, and thus the difference is 1 for this column. Column 2 is a bit more interesting. Here the operation is $0 - 0$, but there is also a borrow indicated from the previous operation. The borrow will reduce the minuend by that amount, and thus the operation is $-1_2 - 0_2$, a negative number. Thus we borrow 10_2 (2 in decimal) from the next column over, and the effective operation is $10_2 + (-1_2) - 0 = 1$. The subtractions in the other columns proceed in an analogous fashion.

All of the possible operations for both subtraction and addition are summarized in Table 1.3 (A and B). In the case of both operations, we are always taking three binary operations and producing two results, one that goes on the carry or borrow bookkeeping line, and one that is the result of the operation. Because there are only three possible binary inputs, there are only 2^3 or 8 possible sets of results. As an example of how to use Table 1.3 (A and B), consider the first column in the previous subtraction example again. The minuend was 0, the subtrahend was 1, and the borrow in (the borrow for this column) was 0. Table 1.3 tells us that the difference is 1, and the borrow out (the number to write in the top row of the next column) is 1. Thus, Table 1.3 allows one to perform any binary addition or subtraction without having to go through the arithmetic logic each time as we did in the previous examples. However, it is more than a mere convenience. As we will discuss in later chapters, Table 1.3 (A and B) allows us to convert arithmetic operations into simple logical ones, effectively permitting the construction of an arithmetic processor from logical gates alone.

TABLE 1.3 Tables for (A) Binary Addition and (B) Subtraction

Addition (A)				
addend 1	addend 2	carry in	sum	carry out
0	0	0	0	0
0	0	1	1	0
0	1	0	1	0
0	1	1	0	1
1	0	0	1	0
1	0	1	0	1
1	1	0	0	1
1	1	1	1	1

Subtraction (B)				
minuend	subtrahend	borrow in	difference	borrow out
0	0	0	0	0
0	0	1	1	1
0	1	0	1	1
0	1	1	0	1
1	0	0	1	0
1	0	1	0	0
1	1	0	0	0
1	1	1	1	1

Binary Multiplication

Moving now from a generalization of second-grade to third-grade arithmetic, let us consider multiplication. Actually, binary multiplication turns out to be easier than grade school multiplication because there will never be a carry when forming the products (see Exercise 1.8 for an example of multiplication in other bases that require carries). This follows from the fact that the ones times table never exceeds 1: $0*0 = 0$, $0*1 = 1$, and $1*1 = 1$. An example illustrates the basic procedure:

```
      0110
    * 1011
      ----
      0110
     0110
    0110
    -------
    1000010
```

As in decimal multiplication, a partial sum is computed for each column by multiplying the digit in the bottom factor with the entire top factor. The sum of these quantities is the final result. Note that forming this sum follows a similar procedure to that given for adding two binary numbers, although in this case multiple numbers may be present.

Negative Numbers in Binary: Signed Magnitude and Two's Complement

We have been implicitly assuming that our mathematical operators take two or more positive numbers and produce a positive number. However, some means to represent negative quantities is also needed. The simplest method, assuming we use only binary numbers, is to let the leftmost bit (most significant) be 0 if the number is positive and 1 if it is negative. For example, assuming we are representing numbers with bytes, 01010111 would be 87 as before, but 11010111 would be –87. The effective range of this convention, known as *signed magnitude*, is 0111111 to 1111111, or 127 to –127, if a single byte is used.

As in any representational system, the quality of the system is judged by how it performs with respect to typical operations on these representations. Signed magnitude, unfortunately, is somewhat wanting in this regard. The following pseudocode shows what a computational system would have to run through in order to produce the correct result when adding two signed magnitude numbers.

```
if (signs same) then
{
    add magnitudes
    give result this sign

}
else    /* signs different */
{
    compare magnitudes
    subtract smaller from larger
    give result sign of the larger
}
```

This algorithm involves a number of conditionals, including the top-level if-then-else, as well as the magnitude comparison within the else clause. These conditionals take time to evaluate, and thus addition and subtraction in this representation is a relatively costly process.

It turns out to be possible to eliminate the conditionals if one uses a representation of negative numbers known as *two's complement.* A negative number in two's complement is simply the difference between 2^n and the positive version of the number, where n is the number of bits in the representation. For example, let us represent –17 in two's complement with one byte. Then +17 would be 00010001, so by construction –17 is 100000000 – 00010001 = 11101111 (that is, $2^8 - 17$). Positive numbers are the same in two's complement as in ordinary binary.

There is a trick for computing the negative of any positive number in two's complement without having to do the subtraction: simply complement all the bits in the positive number and add 1. Thus, for the previous case, we start with +17, or 00010001, flip the bits to yield 11101110, and add 1 to yield 11101111. This works because we want the sum of the positive version of the number and the negative version to be 2^8. Adding the positive version and the complemented version always produces 11111111 and then adding 1 to this always produces 100000000, or 2^8. The range of single byte two's complement numbers is 01111111 (+ 127) to 10000000 (–128). The most significant (leftmost) bit, as in signed magnitude, signals a positive number if it is 0, and a negative number otherwise.

System complement in the Chapter1 folder on the CD-ROM contains a mechanism for computing the two's complement of a negative number. It works similarly to the addition system except that one of the addends (addend1) is computed as the inverse of the topmost row (labeled "original"), and the other addend (addend2) is fixed at 00000001. This, of course, is just our trick algorithm for producing the two's complement. The default value read in with the system is +17, although we can use the system to find other results by clicking on the topmost row of cells.

The significance of this representation is as follows. Suppose we are adding two positive numbers. Positive numbers are unchanged in two's complement; therefore, the addition is as before. Suppose however, we are adding a positive number, +a, to a negative number, –b. By construction, this sum is

$$a + (2^n - b) = (a - b) + 2^n. \tag{1.6}$$

In other words, if we just do ordinary addition, we will get the right-hand side of this equation. But this quantity is simply a–b, assuming we ignore the 2^n, i.e., the bit to the left of the leftmost bit in the byte if it appears. (This will happen if the absolute value of the quantity a is greater than quantity b because then the sum will be greater than 2^n; otherwise, it will be less than 2^n.) Likewise, the addition of two negative numbers –a and –b yields

$$(2^n - a) + (2^n - b) = (-a - b) + 2^{n+1}. \tag{1.7}$$

Again, the right-hand side of this equation shows that we get the correct result if we ignore any bits that appear to the left of the leftmost original bit.

Thus, we have constructed a representation that does not require conditional testing, unlike in signed magnitude. Ordinary addition suffices, once the conversion of negative numbers has been accomplished, and assuming extra higher-order bits are ignored. Here is an example of adding 43 and (–17), or equivalently subtracting 17 from 43, using a single two's complement byte for each.

```
     00101011     (43 in two's complement)
  + 11101111     (–17 in two's complement)
    ---------
🗑 ← 100011010
```

Note that we handle the shaded bit in the ninth position that results from the last carry by sending it straight to the digital trashcan, and thus the answer is 00011010, or 26, as is easily verified by carrying out the same operation in decimal.

As with any arithmetic operation on a finite number of bits, we must keep in mind the possibility of overflow, or a result that cannot be represented by the number of bits in question. In two's complement addition, this will occur when the result of adding two positive numbers is a negative number (the most significant bit in the result ends up as 1), or when the sum of two negative numbers results in a positive number (the most significant bit is 0). For example:

$-113 + (-17)$ (or, equivalently, $-113-17$)

```
  10001111      (–113 in two's complement)
+ 11101111      (–17 in two's complement)
----------
🗑 ← 101111110   overflow!
```

In this case, the sum of two negative numbers is positive because the leftmost bit of the sum is 0 (as usual, we ignore any higher-order bits generated, that is, the shaded bit[1]). Another way of seeing this is that the result should be -130, but the smallest negative number in two's complement for one byte is 10000000, or -128. Whenever there is an overflow, the result is declared meaningless, although we may wish to set a flag in hardware to indicate that the overflow occurred. You may wonder why we do not just add an extra bit so that we can properly represent the cases where overflow occurs. The answer is that we have to stop adding bits somewhere, and wherever we stop there remains the possibility of generating a number that cannot be represented with the current configuration. Furthermore, once multiplication is allowed, not to mention raising a number to a power and factorial, it is very easy to exceed the allotted bits, and so some means of detecting overflow must be in place.

CODES

A code may be thought of as a system of representing either a set of quantities or a set of symbols, within a given base. For example, we have been looking at what are perhaps the most natural binary representations of decimal numbers, but there are alternatives that serve other purposes. Table 1.4 shows three such codes: BCD, Gray, and Parity for the numbers up to 15. We will consider each briefly.

BCD

Computers compute in binary, but people primarily use decimal. This means, for example, that three conversions are necessary for every arithmetic operation: two to take the operands into binary, and one more to take the binary result back into decimal. These conversions are costly, as they use similar algorithms to those that we discussed above. One alternative to pure binary encoding that reduces some of the complexity of the conversion process is BCD (Binary Coded Decimal).

In BCD, each decimal digit is encoded by four binary digits. We know that 4 bits are capable of representing 2^4 or 16 numbers; thus, 4 bits are more than sufficient to code the numbers 0 through 9. Table 1.4 shows that these digits are encoded identically to ordinary binary. When an extra digit appears, this digit is also encoded by 4 bits. For example, 14 is coded by 0001 0100; 97 is 1001 0111. Note that the binary sequences 1010 through 1111 are never used. Thus, BCD is somewhat less space efficient than ordinary binary, although under the right circumstances, this waste is compensated by the ease of conversion.

TABLE 1.4 Three Alternative Binary Codes

Decimal	Binary	BCD	Gray	Parity
0	0000	0000 0000	0000	00000
1	0001	0000 0001	0001	10001
2	0010	0000 0010	0011	10010
3	0011	0000 0011	0010	00011
4	0100	0000 0100	0110	10100
5	0101	0000 0101	0111	00101
6	0110	0000 0110	0101	00110
7	0111	0000 0111	0100	10111
8	1000	0000 1000	1100	11000
9	1001	0000 1001	1101	01001
10	1010	0001 0000	1111	01010
11	1011	0001 0001	1110	11011
12	1100	0001 0010	1010	01100
13	1101	0001 0011	1011	11101
14	1110	0001 0100	1001	11110
15	1111	0001 0101	1000	01111

Gray Coding

Under certain circumstances, it is desirable that each successive number differs from the previous number by only 1 bit. For example, when counting with a CMOS

(Complementary Metal Oxide Semiconductor) chip, a unit of power is used for every bit change. If these changes can be minimized, then power consumption will also be minimized. Gray coding achieves this with the sequence shown in Table 1.4. In the normal binary sequence, moving from 3 to 4 would entail 3 bit changes (the three leftmost bits all change). In Gray coding, in contrast, only the third bit is altered. We will see further use of Gray coding when we discuss Karnaugh maps in Chapter 3.

Parity

Parity coding is a type of coding that enables error detection. It accomplishes this by attaching an extra bit to the code so that the total number of 1s in the string is always even or odd. Table 1.4 shows even parity coding for the regular binary codes in the second column. Every string in the parity column has been adjusted by putting an extra bit in the front to make the number of bits even. For example, the representation of decimal 7 is 10111.

The purpose of this adjustment is to enable a machine receiving these numbers from another machine to check if a possible mistake has been made in transmission. If an odd number of bits is received, then the receiving machine knows that something is wrong because such a string is not legal within the code. This scheme is not perfect because it cannot detect even numbers of errors—the resulting string in this case would still have an even number of bits—but it is far better than having no error detection whatsoever (adding more bits to the error-detecting part of the string can increase the probability of error detection at the cost of reduced transmission speed and greater space to store the strings). If an error is detected, the receiving machine can ask the sending machine to retransmit the code, and this process can continue until an error-free transmission is made.

ASCII and Unicode

Numeric codes can represent characters as well as numbers. The most common character code is known as ASCII (American Standard Code for Information Interchange) and is shown in Table 1.5. The 26 English characters, their capitalizations, a number of punctuation characters, as well as control codes such as ESC are represented in this scheme. The table shows the 7 bit representation, bits b_0 through b_6, for these symbols. Actually, ASCII uses 1 byte or 8 bits, so there are a total of 128 characters, which are not specified by the coding; these extra characters are operating system and/or hardware dependent. Note that the 8 bits in the ASCII code can be represented by two hexadecimal numbers for convenience (recall the section on converting between bases when one is an integer power of the other). For example, "e" is either 01100101_2 or 65_{16}.

TABLE 1.5 The ASCII Character Code

b3b2b1b0	b6b5b4							
	000	001	010	011	100	101	110	111
0000	NUL	DLE	SP	0	@	P	`	p
0001	SOH	DC1	!	1	A	Q	a	q
0010	STX	DC2	"	2	B	R	b	r
0011	ETX	DC3	#	3	C	S	c	s
0100	EOT	DC4	$	4	D	T	d	t
0101	ENQ	NAK	%	5	E	U	e	u
0110	ACK	SYN	&	6	F	V	f	v
0111	BEL	ETB	'	7	G	W	g	w
1000	BS	CAN	(	8	H	X	h	x
1001	HT	EM	)	9	I	Y	i	y
1010	LF	SUB	*	:	J	Z	j	z
1011	VT	ESC	+	;	K	[	k	{
1100	FF	FS	,	<	L	\	l	|
1101	CR	GS	-	=	M	]	m	}
1110	SO	RS	.	>	N	^	n	~
1111	SI	US	/	?	O	_	o	DEL

ASCII has more than enough space to encode English, but not nearly enough to encode characters in all the world's languages. Hence, the introduction of a new standard, Unicode (*http://www.unicode.org*), which consists of up to 4 bytes (8 hex digits). At the time of this writing, Unicode comprises approximately 50 scripts capable of representing 250 languages (a single script may be used by more than one language, as in the case of English and Spanish). The 16-bit, or 2-byte version of Unicode is capable of representing the vast majority of scripts. The Unicode encoding for Kana characters, used to supplement Kanji (Chinese ideographs) in Japanese, is shown in Figure 1.3. Note that the normal representation for the 2-byte Unicode standard is 4 hex digits.

	30A	30B	30C	30D	30E	30F
0	＝ 30A0	グ 30B0	ダ 30C0	バ 30D0	ム 30E0	ヰ 30F0
1	ア 30A1	ケ 30B1	チ 30C1	パ 30D1	メ 30E1	エ 30F1
2	ア 30A2	ゲ 30B2	ヂ 30C2	ヒ 30D2	モ 30E2	ヲ 30F2
3	イ 30A3	コ 30B3	ッ 30C3	ビ 30D3	ャ 30E3	ン 30F3

FIGURE 1.3 The 36 Kana characters and their Unicode 2-byte equivalents.

SUMMARY

Numbers and number systems are an essential component of digital design. First we saw that positional number systems are the most useful for arithmetic operations. Such systems may be in any base—the most useful for digital circuits is binary. Next we discussed arithmetic operations in binary, which are essentially generalizations of the same operations in decimal. However, the representation of negative numbers with an extra bit is something unique to digital design. Then we showed why two's complement is a desirable means of achieving this representation. Finally, the chapter treated codes, or alternative ways to represent a set of quantities or symbols. A recurrent theme introduced in the chapter, and one that will be expounded upon in later chapters, was that arithmetic and other operations on binary quantities can be reduced to logical operations, making them amenable for use by digital circuits.

EXERCISES

1.1 How many unique quantities can five digits in base 3 represent? How many quantities can four digits in base 17 represent?

1.2 College 1 has decided to go "dozenal," and keeps score only in base 12. College 2 has also decided to rebel again the tyranny of base 10, and keeps score only in base 8. The two meet in the final of the division basketball tournament. College 1 scores 74 in their system and college 2 scores 127 in theirs. Who won?

1.3 Convert 458_9 to base 3 without going through base 10.

1.4 Convert $4F_{16}$ to base 8 without going through base 10 (Hint: Go through binary.)

1.5 Convert the following numbers to decimal:
(a) $BA4_{16}$
(b) 555_7
(c) 1101.1101_2

1.6 Convert 823_{10} to the following bases:
(a) base 2
(b) base 8
(c) base 16

1.7 Perform the following additions in binary:
(a) 12 + 17
(b) 82 + 46

1.8 Perform the following multiplications:
(a) $22_3 * 12_3$
(b) $36_7 * 25_7$

1.9 In each case, determine the base b:
(a) $12_b + 34_b = 101_b$
(b) $427_b + 211_b = 640_b$
(c) $22_b + 8_b = 30_b^2$

1.10 Determine all bases for which the following is true:
$(121_b)^{1/2} = 11_b$

1.11 Perform the following in single byte two's complement. Indicate whether an overflow has occurred.
(a) 41 + 25
(b) 73 + 62
(c) 73 + (-31)
(d) −45 + (−22)
(e) −96 − 72

1.12 Conversion from a positive number to its negative counterpart in two's complement entails flipping the bits and adding 1. Conversion in the reverse direction is identical. Why? (Hint: Recall how two's complement numbers are constructed.)

1.13 We discussed how Gray coding involves only 1 bit change between successive numbers. How much more efficient is this coding scheme than ordinary binary in counting up to 15 if efficiency is defined as the total number of bit changes that are needed in the counting sequence?

1.14 How many unique codes are there to represent the quantities 0 to 15 with 4 bits?
1.15 How many hex digits are necessary to represent 4-byte Unicode. How many bits?
1.16 What string does the following ASCII hex sequence represent? 42 6F 6F 6C 65
1.17 01010111 01101000 01101111 00100000 01110111 01110010 01101111 01110100 01100101 00100000 01010111 01100001 01110110 01100101 01110010 01101100 01111001?

LATTICE Exercises

1.18 A local configuration of cells in the *Game of Life* that does not change from iteration to iteration is known as a steady state attractor. Find at least three such attractors (Hint: Try entering a random starting configuration, by left-clicking on the desired cells, and observing local patterns when this system stops evolving.)
1.19 Try adding the numbers 10101010 and 11110000 with the "addition" system. Is the answer the system produces correct? Why or why not?

ENDNOTES

1. Make sure that you don't confuse this bit with the overflow itself. There is another means of detecting overflow using this bit, discussed in Chapter 4, but it is not equivalent to the simple existence of this bit.
2. This is a kind of mathematical joke. Math jokes are thin on the ground, but there are a few out there.
3. One possible answer is Jebediah Cleisbotham (or is that obvious?). Give the other.

2 Boolean Algebra

In This Chapter

- Introduction
- Logical Functions in Boolean Algebra
- Simplification
- Summary
- Exercises

INTRODUCTION

Formal logic can be traced back, as with many seminal areas in Western thought, to the Greek philosophers. Aristotle, in particular, was the first thinker to formalize what had been up to that point, a grab bag of principles. Aristotle was concerned with arguments of the following form, known as syllogisms:

All men are mortal.	(premise 1)
Harry is a man.	(premise 2)
Therefore, Harry is mortal.	(conclusion)

Given that both premises 1 and 2 are the case, we can draw the conclusion that Harry is not one of the gods, and that he will one day perish. The key to understanding logic, and also the reason it can be automated with integrated circuits, is that logical conclusions such as this are drawn by virtue of their form alone. The above argument is still valid if we substitute Morty with Harry, and change the conclusion to "Morty is mortal." Or (with apologies to Noam Chomsky), even if the first premise is the nonsensical "All men are colorless and green," then a conclusion can still be drawn, namely, that Harry is colorless and green.

As it turns out, it is not possible to implement syllogisms like the above with the kinds of circuits that we will consider. In order to express universal claims such as premise 1, it is necessary to use a language known as the predicate calculus, and this formalism is usually implemented in software rather than hardware. Thus, we will be studying a weaker formalism known as Boolean algebra, invented by the 19th-century philosopher and mathematician George Boole. You should note, however, that as with all logic, Boolean algebra is a system that abides by strict mechanical rules, and it is this fact that will eventually allow us to implement it with digital circuits.

LOGICAL FUNCTIONS IN BOOLEAN ALGEBRA

Logical functions are constructed from variables and operators. A variable is any letter or string of letters that stand for a particular proposition. For example, P may stand for the proposition "it is cloudy," or "Tuesdays are happy days." Whatever P stands for, it can take only one of two values, either 1 or 0, corresponding to true and false respectively. This is it. P cannot be half-true, false today and true tomorrow, or contain a grain of truth; it must squarely fall into one of these two categories.

Strictly speaking, logic that uses "true" and "false" is known as Boolean algebra, and logic that uses 1s and 0s is known as switching algebra. The latter is more suited to digital circuits, which can manipulate voltages corresponding to the numeric quantities 1 and 0, but not symbols like "true" and "false." Both forms however, are operationally equivalent (and also both are equivalent to the propositional calculus), which means that an operation in one is equivalent to an operation in another, with the appropriate substitutions of 1 for "true" and 0 for "false."

Operators are objects that state how symbols combine to produce new values of either 0 or 1. This new value is a strict function of the values of the variables and the operator itself. For example, our first operator, "+" corresponds roughly to the way the word "or" is used in English, and yields 1 whenever either or both of its arguments are 1. It is shown in Table 2.1. This table is known as a truth table, and enumerates all the possible combinations of the variable values. In this case, there are two variables, so there are a total of 2^2 or 4 combinations: 00, 01, 10, and 11.

TABLE 2.1 The Truth Table for the "+" Operator

P1	P2	P1 + P2
0	0	0
0	1	1
1	0	1
1	1	1

The operator "·", which corresponds to the way the word "and" is used in English, yields 1 if and only if both of its arguments are 1. The truth table for "·" is given in Table 2.2.

TABLE 2.2 The Truth Table for the "·" Operator

P1	P2	P1 · P2
0	0	0
0	1	0
1	0	0
1	1	1

You can begin to see why this formalism is called "algebra." The "·" operator works identically to multiplication, and the "+" connective works like addition with the exception that 1 + 1 is 1 and not 10, as it would be in binary.

The LATTICE program is designed to automate the representation of truth tables, such as those in Tables 2.1 and 2.2, and also automate the computation of truth values in those tables. Load system OR from the Chapter2 folder on the CD-ROM and then center-click (or right-click if you don't have a center button, although this will change the color of the cell) on the cell labeled "X + Y." The value of this cell is the sum of the variables X and Y, as can be seen on the truth table on the left side of the screen, which is similar to that at the top of Table 2.1. The operator values can be seen under the column labeled "F," for function.

ON THE CD

In general, the program computes not only values in the truth table but also uses these computations to determine the next state of the set of cellular machines (as displayed in the matrix on the right of the window). In this case, the system computes the sum of the cell to the left and to the right of the labeled cell. If either is on, then the labeled cell comes on. Also, if both are on, then the center cell will also come on, in accord with the definition of "+." Try left-clicking on either or both of these cells, and then iterating the system once (click on the single forward arrow below the cells) to see this effect.

The reason this works is that the variables X and Y are defined to be 1 whenever the left and right cells are on, respectively. To see this, click on the labeled cell, and then left-click on any "var" above the truth table. You will see that the trigger for X is left of the current or center cell (left of the dotted cell), and for Y is right of center (right of the dotted cell). Close this box by clicking on cancel. There are also two other machines in this system, one for X and one for Y. Both are defined in such a way as to stay constant as the simulation advances. Clicking on a "var" after clicking on either machine reveals that this works because these single variable machines are looking at their own value in the prior state to determine their next state. This is represented by the fact that the single variable P1 corresponds to the darkened dotted cell (see "Tricks of the Trade" in the Appendix for a more complete description).

In summary, this system contains three machines, two that perpetuate cell contents, and another consisting of a single cell that computes the sum of the cells to the left and right of it (there is also the background machine, which is always present). The AND system works identically except that the key cell only comes on when both of its neighbors are on. Load this system and confirm that this is the case.

"+" and "?" are binary operators, or operators that take two symbols and return a new value. There is another operator—negation—that operates only on a single symbol, which will prove important in the construction of logical functions. Negation, which is also known as complementation, is indicated by "'". As its name suggests, it reverses logical values.

TABLE 2.3 The Truth Table for the "'" or Negation Operator

P	P'
0	1
1	0

A logical function or logical expression is some combination of symbols and possibly one or more of the three operators "+", "·", and "'". Here are some examples of functions: P + Q, R, (B754 · B721) + W23, [(P' + Q') · (Q + R)]'. Note that brackets or parentheses may be necessary to clarify the order of operation. For example, to compute the value of the last function, first we would compute (P' + Q'), then (Q + R), then logically multiply the result together, and then negate that result.

Truth Tables

If, as has been previously claimed, a variable can only take one of two values, 0 or 1, and if operators return one of these two values only, then it should be possible to exhaustively compute all possible values of a function by computing the value of each operator in the function and combining the results. These values are usually listed in a truth table similar to the kind that we have already seen with the three elementary operators. Table 2.4 is an example with a more complex function, [(P1 + P2) · P3]'. In principle, it is possible simply to list the value of this function for each row by keeping track of partial values in one's head, but it is often easier to break the computation into parts. Here we first compute (P1 + P2) from the rule for the addition operator in Table 2.1, then we multiply this with P3 to yield (P1 + P2) · P3 by the rule for the multiplication operator given in Table 2.2. The final result, which is just the complement of the previous column, follows from the rule for the negation operator given in Table 2.3. Note that these calculations are carried out for each combination of values for P1, P2, and P3.

ON THE CD

One simple use of LATTICE is to derive the values of a logical function. We will illustrate this capacity with the function in Table 2.4. After first starting the program on the CD-ROM or choosing New from the File menu, the default function "P1 + P2" appears. We will change this to our desired function in a number of steps. First, click on "add var" at the bottom of the window. There should be three variables on the left: P1, P2, and P3. Next, click on the variable P1 in the fourth column (there is an open parenthesis before it), and use the pull-down menu to select "expand." Now modify the expression by changing operators and variable names using the appropriate pull-down menus until it resembles ((P1 + P2) * P3). The final step is to negate the entire expression by selecting the negation operator ("~") in the "*" operator column (this should be the seventh column). During each step, the values in the table should change in accord with the indicated function, and the values under the "F" column (for function) should be the same as the last column in Table 2.4 when you are through.

TABLE 2.4 The Truth Table for a Complex Function

P1	P2	P3	(P1 + P2)	(P1 + P2) · P3	[(P1 + P2) · P3]'
0	0	0	0	0	1
0	0	1	0	0	1
0	1	0	1	0	1
0	1	1	1	1	0
1	0	0	1	0	1
1	0	1	1	1	0
1	1	0	1	0	1
1	1	1	1	1	0

The following notes apply to all truth tables, regardless of the nature of the logical function:

- In general, if there are n variables, there will be 2^n rows in the corresponding truth table. For example, the function in Table 2.4 has three variables and thus there are 2^3 rows in the truth table, which cover all the possible combinations of the values of the variables.
- Although it is possible in principle to list the possible combinations of the variable values in any order, by convention we list them in counting order. For example, if we think of P1, P2, and P3 in Table 2.4 as the successive digits in a binary number, the rows in the table count from 000 to 111, or 0 to 7 in decimal.
- Extra columns may be inserted as necessary to break down the problem into smaller components that may be more easily computed. These are optional but may help to clarify the computation. The mechanism for doing so in the case of Table 2.4 has already been described.

Tautology, Equivalence, and Logical Laws

If a function is 1 in every case, regardless of the value of the variables, then it is known as a tautology. A simple tautology is P + P', which we can see in Table 2.5.

TABLE 2.5 The Truth Table for a Simple Tautology

P	P'	P + P'
0	1	1
1	0	1

Regardless of whether P is 1 or P is 0, the function P + P' is always 1.

Of special interest in logic are tautologies that involve the equivalence operator. As one might expect, this operator yields 1 only if both of its arguments are the same, that is, both 0 or both 1. The truth table for equivalence shown in Table 2.6, ordinarily represented with the symbol "≡", represents this fact.

TABLE 2.6 The Truth Table for the Equivalence Operator

P1	P2	P1 · P2
0	0	0
0	1	0
1	0	0
1	1	1

If a logical function, F1 is equivalent to another logical function F2, then wherever we see F1 we may substitute F2 or vice versa. Why? Because from a logical point of view, as we have stressed, the values the function takes is the only thing that matters. Therefore, if two functions always take the same values for the same combination of variable values, then there is no effective difference between them.

32 Digital Design: From Gates to Intelligent Machines

This allows us to construct a number of logical laws based on equivalences, which we will then use to simplify expressions. An example of such an equivalence are the functions (P1 + P2)' and (P1' · P2'), or in words, the complement of a sum is equivalent to the product of the complements. The equivalence between these expressions is revealed by the truth table shown in Table 2.7.

TABLE 2.7 The Equivalence Between Two Expressions

P1	P2	(P1 + P2)'	(P1' · P2')	(P1 + P2)' ≡ (P1' · P2')
0	0	1	1	1
0	1	0	0	1
1	0	0	0	1
1	1	0	0	1

Regardless of the values of P1 and P2, (P1 + P2)' always has the same value as (P1' · P2'), as can be seen in the columns corresponding to these expressions. This in turn, implies that the expressions are equivalent, as can be seen in the final column. This particular equivalence is known as DeMorgan's law; it will prove extremely important in the construction of logical circuits.

ON THE CD

System DeMorgan in the Chapter2 folder on the CD-ROM illustrates this law. This system consists of two machines, each representing the respective sides of the equivalence. Clicking in the cells for "expr1" and "expr2" and observing the change in the truth table to the left will confirm this. Also, note that both of these tables contain the same values for the function column, which means by definition they are equivalent. In addition, the two machines produce identical behavior because the variables P1 and P2 are defined identically for both. You can see the definitions for these variables by clicking on "var" in the first two columns above the variables for each machine (they both refer to the same P1 and P2 cells, although the definitions are offset one cell in the vertical direction between the machines because the positions of P1 and P2, relative to the "expr1" and "expr2" cells are different). You can observe the identical behavior by advancing the animation. Regardless of the values of P1 and P2, the cell values for the machines will be identical. This is an alternative means of checking the equivalence of expressions: if two machines refer to the same inputs, and if their behavior is identical, then their corresponding functional definitions must be equivalent.

Other Useful Logical Operators

There are two other operators, NOR, and NAND that will be useful in the construction of digital circuits. Their truth tables are shown below. NAND is just the complement of "·" (the AND operator) and NOR is just the complement of "+" (the OR operator).

TABLE 2.8 The Truth Tables for the NAND and NOR Operators

P	Q	P NAND Q	P	Q	P NOR Q
0	0	1	0	0	1
0	1	1	0	1	0
1	0	1	1	0	0
1	1	0	1	1	0

ON THE CD

The truth table in system NAND, on the CD-ROM, illustrates that the NAND operator is simply a convenient way to express the complement of the product of two variables. This can be seen by the equivalence of the two given expressions (as indicated by the unvarying column of 1s beneath the "≡"). Likewise, system NOR shows the equivalence of this operator and the complement of the sum of two variables. By the way, running either of these systems produces uninteresting results. These functions are always 1, and thus every cell is always on regardless of how P1 and P2 are defined.

Another commonly used operator is exclusive OR, abbreviated XOR (some texts use the "⊕" symbol). The truth table for XOR is shown in Table 2.9. Exclusive OR usually captures the typical use of the word "or" in ordinary discourse better than the "+" operator, which represents inclusive OR. When we say X or Y, in most cases we mean X or Y, but not both. That is, we really mean X XOR Y. For example, if you say to yourself that tomorrow you will either become more serious about digital design *or* you will retire to a Buddhist monastery, you are probably precluding the possibility of doing both, as in becoming a digital designer by virtue of meditation alone.

ON THE CD

System XOR on the CD-ROM contains a single cell in the center defined by this operator, and the cells to the side are constant as in the OR and AND systems. Confirm that this system behaves as desired by setting the X and Y cells by clicking on them (check all four of the combinations shown in the truth table).

TABLE 2.9 The Truth Table for XOR

P1	P2	P1 XOR P2
0	0	0
0	1	1
1	0	1
1	1	0

How many logical operators are there? Assuming the operator acts on two variables, the corresponding truth table will have four rows. Each of these rows can take two possible values, so there are $2*2*2*2 = 2^4$ or 16 possible binary logical operators. By the same reasoning, an operator over n variables will have 2^n rows in its table, and the number of possible outcomes, or sequence of 0s and 1s in the table will then be 2 to this power, or 2^{2^n}. This quantity grows very quickly as a function of n. There are 64 K possible functions of four variables, and $\sim 10^{77}$ possible functions of eight variables.

SIMPLIFICATION

The goal of digital design is not merely to design a circuit that works, but to find the most compact circuit that accomplishes a given task. A more elegant design means a smaller circuit, which usually translates into a circuit that is less expensive. In order to accomplish this, it is often necessary to simplify the logical expression, which the circuit is attempting to realize. What counts as simpler is a difficult matter that we will take up in later chapters; however, for now we will assume that a function is simpler if it contains fewer logical operators, fewer literals (variable instances), or both.

For example, suppose the function is PQ + PR (as in regular algebra, it is not necessary to write "·"; PQ is the same as P·Q). This function contains three operators, two products, one addition, and four literals. An equivalent expression, P(Q + R) contains only two operators, a single addition, a single product, and three literals. It is a simple matter to prove that the two expressions are equivalent using the truth table method.

However, a serious problem remains in general. Suppose that you are given the first expression, but not the second, and asked to find a simpler form of the first function. Obviously, a truth table won't work in this case, because you can't prove the equivalence of two items if you only have the first to work with. What is needed is a simplification method, enabling the transformation of the first function into the second. We now consider two such techniques, the first based on logical laws, and a second more systematic visual method.

Simplification with Logical Laws

We have already mentioned the fact that if two logical expressions are equivalent, then one can always be substituted for another. This fact paves the way for the first simplification technique we will consider. The general idea is to perform successive substitution on an original expression, with the aid of a set of canonical theorems, until a simpler expression results. The set of theorems we will use for the purposes of this section is listed in Table 2.10.

TABLE 2.10 Ten Theorems of Boolean Algebra and Their Duals

Theorem	Dual
(T1) $X + 0 \equiv X$	(T1)^d $X \cdot 1 \equiv X$
(T2) $X + 1 \equiv 1$	(T2)^d $X \cdot 0 \equiv 0$
(T3) $X + X \equiv X$	(T3)^d $XX \equiv X$
(T4) $X + X' \equiv 1$	(T4)^d $XX' \equiv 0$
(T5) $(X')' \equiv X$	
(T6) $X + Y \equiv Y + X$	(T6)^d $XY \equiv YX$
(T7) $(X + Y)' \equiv (X' * Y')$	(T7)^d $(XY)' \equiv (X' + Y')$
(T8) $(X + Y) + Z \equiv X + (Y + Z)$	(T8)^d $(XY) Z \equiv X(YZ)$
(T9) $XY + XZ \equiv X(Y + Z)$	(T9)^d $(X + Y) (X + Z) \equiv X + YZ$
(T10) $X + XY \equiv X$	(T10)^d $X(X + Y) \equiv X$

Before describing how to use these theorems, it is necessary to describe a number of salient facts regarding this table.

1. A number of theorems use the constants 0 and 1 in addition to variables. One can think of these constants as variables that always take the indicated value.

2. All of the equivalences in the table can be proved by enumerating all possible variable values, as in a truth table. For example, it is easy to see that (T2) is a valid equivalence. If X is 0, then the left-hand side of (T2) is just 0 + 1, or 1 (from the truth table for "+"). Alternatively, if X is 1, then the left-hand side is 1 + 1, which also equals 1. X must be either 0 or 1, hence X + 1 always equals 1.
3. Every theorem on the left of the table is accompanied by its dual on the right. A dual theorem is one in which every 0 has been substituted for a 1 and vice versa, and every "+" has been substituted with "·" and vice versa. For example, in (T4) both occurrences of the 1 are replaced with 0 and the "+" turned into a "·" to produce (T4)^d. It is a metatheorem of the propositional calculus, which we will not prove here: that if an equivalence is a tautology, its dual is also a tautology. This means that we get one or the other side for free. For example, we just proved that (T2) is tautologous; this means, without further argument, that (T2)^d is as well. Note that (T5) has no dual because there are no symbols (0, 1, "+", "·") to be changed.
4. Strictly speaking, the variables in Table 2.1 are metasymbols, not ordinary variables. What this means in practical terms is that they can stand for arbitrary expressions. For example, we know from (T3) that P(Q + R') + P(Q + R') is equivalent to P(Q + R'). Why? This follows immediately if we let X stand for P(Q + R') .
5. A number of theorems are identical to those in ordinary algebra. In particular, (T8) and (T8)^d are the associative laws, and (T9) is the distributive law of numerical algebra. However, (T9)^d is unique to Boolean algebra.

Let us now take a look at a few examples of these theorems at work. Suppose you are given the function PQ + PQ' and asked to find a simpler form. The steps are as follows:

Step	Justification
1. PQ + PQ'	given
2. P(Q + Q')	(T9)
3. P(1)	(T4)
4. P	(T1)^d

First we factor the given function in accord with (T9). This yields (Q + Q'), which can be simplified by (T4) to 1. Any expression multiplied with 1 is just that function, by (T1)^d, and the original function has been simplified as far as it can go. The conclusion of our simplification is that the original expression, despite appearances, is the digital equivalent of a pet rock. It is equivalent to just P, or the no operation function. Incidentally, whenever a function has been simplified, we implicitly create two new theorems. The first is the equivalence between the original function and the resulting simplification, in this case (PQ + PQ') ≡ P. The other is the dual, which comes for free as usual. In this case it is [(P + Q)(P + Q')] ≡ P.

A somewhat more difficult function to simplify is P1P2 + P1?P3 + P2P3. Here is the simplification sequence:

Step	Justification
1. P1P2 + P1'P3 + P2P3	given
2. P1P2 + P1'P2 + P2P3(1)	(T1)^d
3. P1P2 + P1'P3 + P2P3(P1 + P1')	(T4)
4. P1P2 + P1'P3 + P1P2P3 + P1'P2P3	(T9), (T6)^d
5. P1P2 + P1P2P3 + P1'P3 + P1'P2P3	(T6)
6. P1P2(1 + P3) + P1'P3(1 + P2)	(T9), two applications
7. P1P2(1) + P1'P3(1)	(T2)
8. P1P2 + P1'P3	(T1)^d

This simplification illustrates that it is not always obvious which theorem to apply. For example, one might want to factor the given function directly, but this does not lead in a productive direction. It also illustrates that sometimes one must take the counterintuitive step of making an expression longer before simplifying it. For example, Steps 2 and 3 add extra operators and extra variables to the original expression. These difficulties motivate the introduction of a more systematic simplification method, which we introduce in the next two sections.

ON THE CD

It is always possible to check the result of a simplification with a truth table. For example, P1P2P3 + P1P2P3' simplifies to P1P2. This type of simplification, in which a variable gets eliminated, will prove important in our discussion of Karnaugh maps. System simplification on the CD-ROM shows that the original function and its simplification are indeed equivalent (all 1s appear in the "F" column). In order to see the entire function, you will need to expand the truth table window by clicking on the small right arrow in the junction between the truth table window and the cell window.

Minterms and Maxterms

Before describing the next simplification technique, the concept of minterms and maxterms, and the corresponding notions of a sum of product (SOP) and product of sum (POS) representation of a function must be introduced. Table 2.11 shows the minterms and the maxterms for each of the $2^4 = 16$ possible sets of values for four variables. The minterms consist of the products of all of the variables, with the variable complemented (negated) if the corresponding value is 0, and not complemented otherwise. The maxterms consist of sums of the variables, with the variable complemented if the corresponding value is 1, and not complemented otherwise. For example, the minterm for the seventh row, in which the value string is 0111, is W'XYZ, and in the maxterm, W + X' + Y' + Z', the complementation is reversed.

TABLE 2.11 The Minterms and Maxterms for a Function of Four Variables

#	W	X	Y	Z	Minterm	Maxterm
0	0	0	0	0	W'X'Y'Z'	W + X + Y + Z
1	0	0	0	1	W'X'Y'Z	W + X + Y + Z'
2	0	0	1	0	W'X'YZ'	W + X + Y' + Z
3	0	0	1	1	W'X'YZ	W + X + Y' + Z'
4	0	1	0	0	W'XY'Z'	W + X' + Y + Z
5	0	1	0	1	W'XY'Z	W + X' + Y + Z'
6	0	1	1	0	W'XYZ'	W + X' + Y' + Z
7	0	1	1	1	W'XYZ	W + X' + Y' + Z'
8	1	0	0	0	WX'Y'Z'	W' + X + Y + Z
9	1	0	0	1	WX'Y'Z	W' + X + Y + Z'
10	1	0	1	0	WX'YZ'	W' + X + Y' + Z
11	1	0	1	1	WX'YZ	W' + X + Y' + Z'
12	1	1	0	0	WXY'Z'	W' + X' + Y + Z
13	1	1	0	1	WXY'Z	W' + X' + Y + Z'
14	1	1	1	0	WXYZ'	W' + X' + Y' + Z
15	1	1	1	1	WXYZ	W' + X' + Y' + Z'

As we will show shortly, any function can be rewritten as follows:

- the sum of the minterms (i.e., a SOP), wherever the function is 1
- the product of the maxterms (i.e., a POS), wherever the function is 0

Before demonstrating these claims, we will illustrate them with an example. Suppose the function F is Y'(X + W) + Z'. The truth table for this function is given in Table 2.12. The function is 1 for rows 0, 2, 4, 5, 6, 8, 9, 10, 12, 13, and 14, and 0 otherwise. By the previously stated principles, we can rerepresent F as:

1. a sum of the minterms where F is 1:

    F = W'X'Y'Z' + W'X'YZ' + W'XY'Z' + W'XY'Z + W'XYZ' + WX'Y'Z' + WX'Y'Z' + WX'YZ' + WXY'Z' + WXY'Z + WXYZ'

or

2. a sum of the maxterms where F is 0:

    F = (W + X + Y + Z')(W + X + Y' + Z')(W + X'+Y' + Z')(W' + X + Y' + Z') (W' + X' + Y' + Z').

TABLE 2.12 The Truth Table for a Sample Function

#	W	X	Y	Z	F = Y' (X + W) + Z'
0	0	0	0	0	1
1	0	0	0	1	0
2	0	0	1	0	1
3	0	0	1	1	0
4	0	1	0	0	1
5	0	1	0	1	1
6	0	1	1	0	1
7	0	1	1	1	0
8	1	0	0	0	1
9	1	0	0	1	1
10	1	0	1	0	1
11	1	0	1	1	0
12	1	1	0	0	1
13	1	1	0	1	1
14	1	1	1	0	1
15	1	1	1	1	0

These representations are a little clumsy because they involve many terms. For this reason, we use this shorthand:

1. SOP: $F = \Sigma_{WXYZ}(0, 2, 4, 5, 6, 8, 9, 10, 12, 13, 14)$
2. POS: $F = \Pi_{WXYZ}(1, 3, 7, 11, 15)$.

The shorthand works by simply listing the indices of the minterms or maxterms, and prefacing them with a Σ, for the sum in the case of the SOP, and a Π, for the product in the case of the POS. However, they are completely equivalent to the long forms of the SOP and POS expressions. Note that the list of indices for the SOP and POS complement each other; that is, taken together, they comprise every row in the truth table.

Why does the sum of the minterms, wherever the function is 1, represent that function? Because for each combination of truth values where that function is 1, precisely one minterm, and no others, will be 1. This is just the minterm for that row. For example, the previous function is 1 in row 2. This means that we want it to be 1 whenever W is 0, X is 0, Y is 1, and Z is 0. But, the minterm for that row, W'X'YZ', is 1 whenever this combination of values is the case, and no other minterms are 1 for this set of values. Similarly, the other rows where the function is 1 are represented by the minterms for these rows, and the sum (OR) of all of these minterms represents all the places where the function is 1.

Why does the product of the maxterms also represent the same function? This is somewhat more difficult to show, but can be carried out in two steps:

1. $[\Sigma_{WXYZ}(S)]' \equiv \Sigma_{WXYZ}(S')$

Here, the symbol S stands for the set of indices of the minterms. The symbol S' stands for the complement of this set, that is, every index that is not in this original set. In ordinary language, this claim states that the complement of a sum of minterms is equivalent to the sum of the complementary set of minterms. For example, in the previous example, this step would be written as $[\Sigma_{WXYZ}(0, 2, 4, 5, 6, 8, 9, 10, 12, 13, 14)]' \equiv \Sigma_{WXYZ}(1, 3, 7, 11, 15)$. The justification for this is that complementing a function is equivalent to making a function 1 whenever it was previously 0 and vice versa. The sum of the minterms of the complement of the original set thus represents the negation of the function.

2. $[\Sigma_{WXYZ}(S)]'' \equiv \Sigma_{WXYZ}(S')'$.

This follows directly by complementing both sides of what was just shown in Step 1. (Exercise 2.5 asks you to show that complementing both sides of an equivalence maintains the equivalence). However, by (T5) in Table 2.10, the left-hand side $[\Sigma_{WXYZ}(S)]'' \equiv \Sigma_{WXYZ}(S)$, and as we will shortly demonstrate, the right-hand side $\Sigma_{WXYZ}(S')' \equiv \Pi_{WXYZ}(S')$. Therefore, $\Sigma_{WXYZ}(S) \equiv \Pi_{WXYZ}(S')$, which is what we need to show: the left-hand side of this equation is just the SOP representation of a function, and the right-hand side is just the POS representation of a function.

It remains to show that $\Sigma_{WXYZ}(S')' \equiv \Pi_{WXYZ}(S')$, or equivalently, that $\Sigma_{WXYZ}(S)' \equiv \Pi_{WXYZ}(S)$ (if we show this for all S, we also show it for S'). This turns out to be a consequence of the double application of a generalized form of (T7), DeMorgan's law. We will not prove this here but rather illustrate it with a characteristic example. Suppose a two-variable function has as its SOP form $\Sigma_{P1P2}(1,3)$ = P1'P2 + P1P2.

Then

$$\begin{aligned}
\Sigma_{P1P2}(1,3)' &\equiv [P1'P2 + P1P2]' \\
&\equiv (P1'P2) \cdot (P1P2)' &&\text{by (T7)} \\
&\equiv (P1 + P2') \cdot (P1' + P2') &&\text{by (T7)}^d\text{, applied to both terms} \\
&\equiv \Pi_{P1P2}(1, 3)
\end{aligned}$$

The final step can be verified by reference to Table 2.13, which shows the minterms and maxterms for a two-variable function. A more general proof would follow the same procedure of applying DeMorgan's law to the sum, and then leverage the fact that the minterm and maxterms in a given row are complements of each other (you are asked to show this in Exercise 2.7).

TABLE 2.13   Minterms and Maxterms for a Two-Variable Function

#	P1	P2	Minterm	Maxterm
0	0	0	P1'P2'	P1 + P2
1	0	1	P1'P2	P1 + P2'
2	1	0	P1P2'	P1' + P2
3	1	1	P1P2	P1' + P2'

ON THE CD

System minmax in the Chapter2 folder on the CD-ROM demonstrates that

$$\Sigma_{P1P2}(1,3)' \equiv [P1'P2 + P1P2]' \equiv (P1 + P2')(P1' + P2') \equiv \Pi_{P1P2}(1,3).$$

Exercise 2.13 asks you to demonstrate with LATTICE that $\Sigma_{P1P2}(1,3) \equiv [P1'P2 + P1P2] \equiv (P1 + P2)(P1' + P2) \equiv \Pi_{P1P2}(0,2)$.

It is not necessary to understand every aspect of this proof to be able to use it. The important thing for now, is to understand the import of the proof. Each function has two representations: one the sum of the minterms where the function is 1, and the product of the maxterms where the function is 0.

NOTE

We know that the SOP and POS characterizations can represent any function. This implies that the three basic operators, AND, OR, and NEGATION, are also sufficient to represent an arbitrary function—there is no need to use operators such as XOR and NAND if we choose to ignore them.

The question remains as to what is the minimum number of operators that are required for this task. As it turns out, the answer is one: NAND. This can be shown in two steps. First, we can represent an arbitrary SOP with only positive literals with NAND. For example, AB + CD ≡ (A NAND B) NAND (C NAND D), as Table 2.14 shows. Likewise, SOPs with more product terms or longer products can also be represented with just NANDs by simply expanding the number of clauses or the number of variables in each clause. What about a negated variable? It is easy to show that A NAND A is equivalent to A. (try proving this with a truth table). Thus, only NANDs are absolutely necessary to represent any function. For example, AC' + A'B + BC is equivalent to (A NAND (C NAND C)) NAND ((A NAND A) NAND B) NAND (B NAND C). The conversion of an expression to one using only NAND gates or NAND gates and inverters will prove important in the construction of combinational circuits, as we will show in Chapter 3.

Karnaugh Maps and Minimization

Once the minterms of a function are known, it is ready to be minimized (that is, reduced to the least complex SOP or POS expression). This can be done algebraically, but is much easier to grasp it [remove it] visually on what we call a Karnaugh map. The steps for minimizing a function on a Karnaugh map are as follows:

1. Enter the minterms for the function on the map.
2. Determine all the prime implicants of the function.

TABLE 2.14 The Equivalence of a Function and its NAND Characterization

A	B	C	D	AB + CD	A NAND B	C NAND D	(A NAND B) NAND (C NAND D)
0	0	0	0	0	1	1	0
0	0	0	1	0	1	1	0
0	0	1	0	0	1	1	0
0	0	1	1	1	1	0	1
0	1	0	0	0	1	1	0
0	1	0	1	0	1	1	0
0	1	1	0	0	1	1	0
0	1	1	1	1	1	0	1
1	0	0	0	0	1	1	0
1	0	0	1	0	1	1	0
1	0	1	0	0	1	1	0
1	0	1	1	1	1	0	1
1	1	0	0	1	0	1	1
1	1	0	1	1	0	1	1
1	1	1	0	1	0	1	1
1	1	1	1	1	0	0	1

3. Determine the smallest subset of these prime implicants that cover all the prime implicants.
4. Name these implicants and construct the final expression.

We now consider each step individually:

1. Entering the minterms on the map

The minterms for an expression can be computed as in the last section. For example, if the expression is W'XYZ + WXY' + WXY + X'YZ + W'X'Y'Z', then the minterm SOP representation is $\Sigma_{WXYZ}(0, 7, 8, 9, 12, 13, 14, 15)$, as can be readily verified with a truth table. Figure 2.1(A) shows a Karnaugh map for four variables, and how the minterms for this function are entered on the map.

The map has $16 = 2^4$ squares, with each corresponding to the numbered minterm. A three-variable function would correspond to a map of $8 = 2^3$ squares and a five-variable function to a map with $32 = 2^5$ squares, although generally, maps with five or more variables are unwieldy to work with by hand. We place 1s wherever the function has the value 1, and leave the other entries blank. Note that each numbered minterm corresponds to the definition originally given in Table 2.11. For example, minterm 7 in Figure 2.1(A) corresponds to the case where $W = 0$, $X = 1$, $Y = 1$, and $Z = 1$, or W'XYZ.

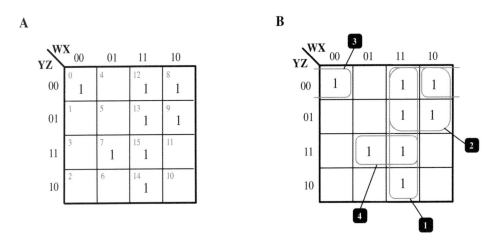

FIGURE 2.1 (A) A Karnaugh map for four variables with minterms indicated, and (B) the prime implicants for these minterms.

Note that the numbering is out of order; for example, the sequence for the first column is 0, 1, 3, 2. This is a consequence of the fact that the coding for each of the two variables WX along the top, and YZ along the side, is the Gray coding and not the usual numerical ordering. Recall from Chapter 1 that the defining characteristic of Gray coding is that each successive number differs from its predecessor by 1 bit only. The reason for this coding scheme will soon become apparent.

2. Determining the prime implicants (PIs)

The second step consists of finding all the prime implicants of the function. A prime implicant has the following characteristics on a Karnaugh map:

- It consists of 2^n minterms, where n is an integer. In the case of a four-variable map, this means that each prime implicant will have either 1, 2, 4, 8, or 16 minterms (in the latter case, all the cells in the map will be 1s).
- It forms a contiguous rectangle, where continuity is defined over the three dimensional torus (doughnut-shape) formed by joining the edges of the map. This requires an explanation. In Figure 2.1, minterms 0 and 4 are contiguous, but so are 0 and 8, because the Karnaugh map, as we have drawn it on the flat page in Figure 2.1(A), is only an approximation to the "real" map. The real map is actually a torus. You can imagine the torus formed by first joining the left edge to the right edge, forming a cylinder, and then joining the top of the cylinder to the bottom. When this is done, the adjacency relations are as shown in Figure 2.2, which shows the torus viewed from the top and the bottom. Note for example, that minterm 2, which appears to be adjacent only to minterms 3 and 6 on the flat projection of the map in Figure 2.1(A), is also adjacent to minterms 10 and 0 in Figure 2.2 (the latter relation can be seen if you imagine starting at minterm 2 and moving through the hole in the doughnut, which brings you immediately to minterm 0).
- It is not subsumed by a larger implicant. If there are 2^n terms in the form of a contiguous rectangle, but they are completely "swallowed" by a larger set, then it does not count as a prime implicant.

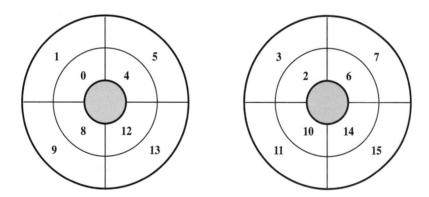

FIGURE 2.2 Top and bottom views of the "Karnaugh torus."

Figure 2.1(B) shows the four PIs for the map in Figure 2.1(A).

1. PI 1 comprising minterms 12, 13, 14, 15
2. PI 2 comprising minterms 8, 9, 12, 13
3. PI 3 comprising minterms 0, 8
4. PI 4 comprising minterms 7, 15

Note that PI 3 wraps around the edges because minterms 0 and 8 are adjacent, as previously explained. Note also that cells 8 and 12 do *not* constitute a prime implicant by themselves. These cells are completely subsumed by a larger implicant, PI 2. Finally, note that prime implicants are allowed to overlap; this occurs for example, in the case of PIs 1 and 2, 2 and 3, and 1 and 4.

3. Finding a minimal set of prime implicants that cover all the 1s

To start this step, we find what we call the essential prime implicants. A PI is essential if it contains a minterm that no other PI contains. If this is the case, and the PI is not to be included in the final set, then this minterm would not be covered. Hence, these PIs are essential. In Figure 2.1, every PI is essential. PI 1 uniquely covers minterm 14, PI 2 uniquely covers minterm 9, PI 3 uniquely covers minterm 0, and PI 4 uniquely covers minterm 7. Therefore, each one is needed. Furthermore, taken together, the PIs cover every minterm. Therefore, PIs 1-4 constitute a minimal set.

The justification for forming PIs is that each will correspond to a product term, with the larger the number of minterms in the PI, the smaller the number of variables in the product. For example, let us consider PI 4. This covers minterms 7 and 15 and can also be represented in functional form: W'XYZ + WXYZ. However, this can be factored using (T9) in Table 2.1 to XYZ(W + W'). But W + W' is just 1, so PI 4 is just XYZ. Note that minterms 7 and 15 differ in just one variable: W. This is what permits the factoring out and elimination of this variable and the justification for using Gray coding in the Karnaugh map. If adjacent minterms differed by more than one variable, this type of factoring would not be possible.

ON THE CD

On the CD-ROM, system simplification, which we have already seen in the context of a prior section, contains a truth table illustrating an analogous equivalence:

$$(P1P2P3 + P1P2P3') \text{ and } P1P2.$$

What about PIs 1 through 3, each covering four minterms? These can be reduced to a single product term with two variables, such that the variables are those that have the same value for all the minterms. This can be shown with a series of simplifications in the case of PI 1; the other cases are analogous:

$$\begin{aligned}
WXY'Z' + WXY'Z + WXYZ' + WXY'Z' &\equiv WX(Y'Z' + Y'Z + YZ' + Y'Z') \\
&\equiv WX(Y'(Z' + Z) + Y(Z' + Z)) \\
&\equiv WX(Y'(1) + Y(1)) \\
&\equiv WX(Y'(1) + Y(1)) \\
&\equiv WX(Y' + Y) \\
&\equiv WX(1) \\
&\equiv WX
\end{aligned}$$

4. In general, the naming rule for PIs is as follows: The name of the PI will correspond to a product of the variables that don't change, with the variable complemented if it is always 0, and not complemented otherwise.

By this rule, PI 2 is WY' and PI 3 is X'Y'Z'. The final SOP expression for the function is the sum of PIs 1 through 4, or WX + WY' + X'Y'Z' + XYZ. Compare this to the original function, W'XYZ + WXY' + WXY + X'YZ + W'X'Y'Z'. Five product terms have been reduced to four, and the total number of literals, or variable instances, has been reduced from 17 to 10.

Here is another example. Suppose the truth table for a function reveals that it can be represented as SWXYZ(0, 1, 2, 3, 4, 5, 6, 14, 15) (Figure 2.3(A)). There are five PIs for this function (Figure 2.3(B)).

1. W'X' comprising minterms 0, 1, 2, 3
2. W'Y' comprising minterms 0, 1, 4, 5
3. W'Z' comprising minterms 0, 2, 4, 6
4. XYZ' comprising minterms 6, 14
5. WXY comprising minterms 14, 15

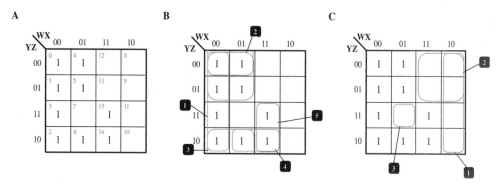

FIGURE 2.3 The minterms corresponding to the function SWXYZ(0, 1, 2, 3, 4, 5, 6, 14, 15).

All of the PIs are essential except for 4, which contains minterms covered by PIs 3 and 5. The essential PIs cover every minterm (in the case where this does not happen, one simply needs to choose a minimal set of PIs to cover the minterms that have not yet been covered; see Exercise 2.8). Thus, the function is characterized by the minimal SOP W'X' + W'Y' + W'Z' + WXY.

Is this the simplest characterization of the function? The answer is not necessarily. The reason for our hesitation is that we must always check at the very least the POS representation also. Recall that every function can be written as a sum of its minterms or as a product of its maxterms. Likewise, every function has a minimal POS characterization in addition to its minimal SOP expression. In order to determine this expression, the same minimization procedure can be used, but this time we will cover the 0s rather than the 1s, which by construction, are the empty minterms in a table.

For example, Figure 2.3(C) shows the PIs for the POS representation:

1. W' + X, comprising maxterms 8, 9, 10, 11
2. W' + Y, comprising maxterms 8, 9, 12, 13
3. W + X' + Y' + Z', comprising maxterm 7

The PIs are sums, of course, instead of products, because the target representation is a product of these sums, and variable complementation is opposite to the prior procedure. If a variable is 1 for every maxterm, then the variable is complemented, and is not complemented otherwise.

Every PI in Figure 2.3(C) is essential and all maxterms are covered, so the final POS representation is (W' + X)(W' + Y)(W + X' + Y' + Z'). This is slightly smaller than the minimized POS representation, W'X" + W'Y' + W'Z' + WXY, in that it has three products as opposed to four sums, and has a total of eight variable instances as opposed to nine. Thus, it is the preferred minimal representation in this case.

NOTE

Are we there yet? That is, does it suffice to examine both the minimal SOP and minimal POS expressions to find the absolute simplest representation of a function? Unfortunately, there are two limitations of this procedure. First, as the number of variables grows much larger than 10 or so, even computer implementations of the minimization algorithm (you will normally use a piece of software to minimize a large function rather than doing it by hand) may not be able to find the absolute minimum form. The approximations the program returns may be good enough in the sense that they have considerably reduced the size of the function, but they will not necessarily be the best SOP or POS representation.

There is another and more profound limitation on the previously described procedure. Neither the SOP nor the POS target representations may be the ideal target representation for a function. That is, there may be alternative representations that turn out to be simpler.

Consider for example, the Karnaugh map shown in Figure 2.4. It is not hard to see that either a POS or SOP representation will have eight constituents, as there is no way to collect any of the minterms or maxterms into PIs larger than a single cell. However, there is an obvious symmetry to the checkerboard pattern, and therefore there ought to be a way to describe this pattern with a simple characterization. In fact, an alternative target representation consisting of XORs cleanly captures this symmetry. It is not difficult to show that (W XOR X) XOR (Y XOR Z) describes the function, and what would have been an eight term function has thus been reduced to a single term.

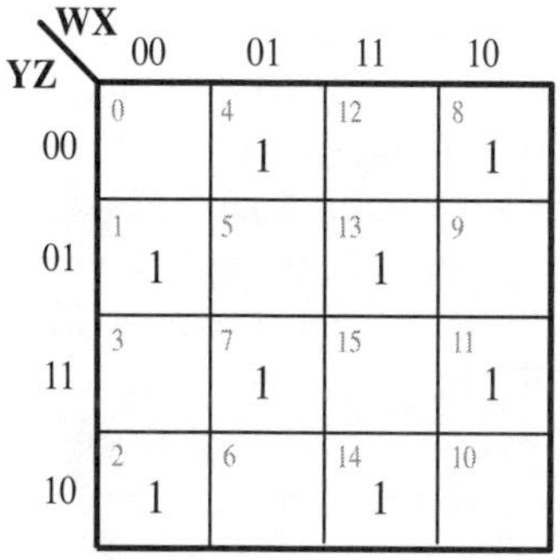

YZ\WX	00	01	11	10
00	0	4 1	12	8 1
01	1 1	5	13 1	9
11	3	7 1	15	11 1
10	2 1	6	14 1	10

FIGURE 2.4 A Karnaugh map in which the minterms produce a checkerboard pattern.

But the lessons to be learned from this pattern do not stop there. Notice that the sum of the positive (not complemented) variables for every minterm is an odd number. If we could devise a counting machine generating a 1 if and only if the sum was odd, we would therefore be able to characterize this pattern. Moreover, this machine would be able to detect the existence of a checkerboard pattern not only for four variables, but for any number of variables. In Chapter 5 we will create such a machine and further compare it with the combinational SOP and XOR realizations of the same function.

In summary, there are two important lessons here. The first and more specific lesson concerns the nature of representation. If it is extremely difficult or costly to capture a pattern with a given target representation, this usually implies that there is an alternative representation in which it is simple and cheap to capture that pattern. Why? The pattern must contain a degree of regularity to elude the original target representation. This very regularity that made it costly to capture in the first place, can be leveraged to find a simpler description, as in the checkerboard case.

The second and more abstract lesson is that digital design may appear at times a mechanical if somewhat laborious exercise. However, when all the possibilities are truly considered, it becomes more of an art, as in any true design process. We will return to this topic throughout the book.

ON THE CD

As masters of Boolean algebra, we are now in a better position to understand the first LATTICE example, the *Game of Life* (system life in Chapter1 on the CD-ROM). The game consists of the sum of two variables: P1 and P2. This means that if either P1 or P2 is true, then the cell will become or stay alive; otherwise it will die. Each of these variables contains a definition, which can be accessed by clicking on any "var" in the truth table. As the dialog box shows, P1 is true if and only if three of the cells surrounding the cell in question are alive. P2 is true if and only if three or four of the surrounding cells, including the cell being processed, are alive. In addition, the "center necessary" box is checked. This means that an additional condition for P2 is that the cell in question is already alive. These conditions are the definition of the game as described in Chapter 1. Remember also that the expression applies to every cell in the grid simultaneously, and that the new state of the cells is computed as a function of the previous states.

SUMMARY

This chapter introduced the notion of Boolean algebra, which consists of a set of operators acting on variables that can only take the values 0 or 1 and which return the values 0 or 1. This fact allows us to construct enumerative truth tables that show the value of a function for every combination of variable values. Truth tables can also be used to show that one function is equivalent to another. Equivalence relations, in turn, can be used to simplify logical functions. A more systematic simplification procedure involves the determination of the minimal sum of the prime implicants for a function. The general procedure is illustrated in Figure 2.5. The truth table of a function is first formed in order to determine the minterms. These minterms are then entered on a Karnaugh map. Typically, two minimal sums of PIs are extracted from the map, one representing the minimal SOP, and one the minimal POS. From these, the simplest expression is chosen. As remarked at the end of this chapter, this is not necessarily the shortest description of the original function. It may be that an alternative target representation involving different logical operators or even a procedural method captures the function more efficiently.

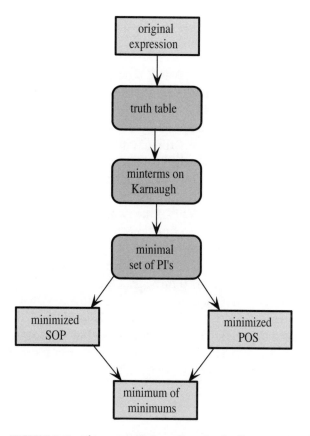

FIGURE 2.5 The general procedure for finding a minimal description of a function. Functions are indicated by rectangles and operations by rounded rectangles.

EXERCISES

2.1 Show the truth tables for the following functions:
   (a) PQ'
   (b) P + Q + R
   (c) [WX + YZ] NAND [W'Z' + WY']

2.2 Show that the following are tautologies with a truth table:
   (a) $X + X' \equiv 1$
   (b) $(X + Y)(X + Z) \equiv X + YZ$
   (c) $XY + X'Z + YZ \equiv XY + X'Z$

2.3 Simplify the following expressions:
(a) P(P + Q)
(b) X + XY + XZ
(c) (X + Y) (X' + Z)(W + Z')

2.4 Construct SOP and POS expressions corresponding to the functions in 2.1.

2.5 Suppose that P1 ≡ P2. Show that it is also the case that P1' ≡ P2'.

2.6 Two light switches are designed such that when both are in the off position the light is off, and any flip of either switch thereafter puts the light in the opposite state. What logical function can be used to achieve this?

2.7 Show that the minterms and maxterms of a given row in a truth table are complementary, regardless of the number of variables involved. (Hint: Apply DeMorgan's law to one of the two, and then show equivalence.)

2.8 Convert the following functions to ones using only NAND gates.
(a) XYZ' + WX + W"Z
(b) AB XOR B'C (Hint: First form a SOP.)

2.9 Show that minterms 0 through 7 form a single product term. (Hint: Use the same factoring method that we used to show that two and four contiguous minterms in a rectangle form a single product.)

2.10 Find a minimal SOP for the following function: W'X'Z + W'X'Y'Z' + W'X'YZ + W'XY' + XYZ + WXY (Hint: Use one or more existing PIs to cover any minterms left after covering with essential PIs.)

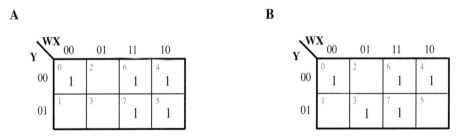

FIGURE 2.6 The three-variable Karnaugh maps for Exercise 2.11.

2.11 Use the same procedure for minimizing a four-variable function on both of the three-variable maps shown in Figure 2.6 (the three-variable map is also a torus).

2.12 Find minimal SOPs and POSs for the following function: SWXYZ(0, 2, 5, 6, 7, 8, 10, 13, 14, 15). Which one is simpler? (Hint: Don't forget that the four corners form a PI.)

LATTICE Exercises

2.13 Show that exclusive OR and equivalence are opposite operators by showing that (P1 XOR P2) _ (P1 ≡ P2) is a contradiction with LATTICE (a contradiction is defined as a function that is always 0). You will need to expand the original P1 and P2 variables in the expression P1 + P2 by selecting this option from the pull-down menu. Then alter the variable names and the operators as necessary.

2.14 Show with LATTICE that SP1P2(1,3) ≡ [P1'P2 + P1P2] ≡ (P1 + P2)(P1' + P2) ≡ PP1P2(0,2).

2.15 Verify the following laws with LATTICE by showing that the equivalence is a tautology.
   (a) T8
   (b) T9d
   (c) T7d

2.16 Verify the following laws with LATTICE by showing that the two machines governed by either side of the equivalence behave identically, as in the system DeMorgan.
   (a) T8
   (b) T9d
   (c) T7d

3 Elementary Combinational Circuits

In This Chapter

- Introduction
- Logic, Gates, and Circuits
- Gates and Integrated Circuits in Practice
- Summary
- Exercises

INTRODUCTION

This chapter is divided into two sections. In the first, we treat digital circuits as the pure manifestation of the abstract world of Boolean logic, without being unduly concerned about the physical implementation of these circuits. In the second, we refine this view by delving into a number of issues that arise in the actual construction of gates and circuits. The emphasis will be on the capabilities of CMOS integrated circuits, the most common form of digital circuitry in use today. Understanding most of the rest of the book will depend on understanding the first part of this chapter. However, in the real world, design is not merely a matter of meeting the logical constraints of the problem at hand, but rather doing so in a cost-conscious and efficient manner. This can be fully accomplished only by exploiting the internal operation of gates in addition to their external, behavioral properties. Thus, the concluding section of this chapter is provided as a way to introduce interested students to the kinds of topics that would arise if they were to become full-fledged professional digital designers.

LOGIC, GATES, AND CIRCUITS

The topic of this section is combinational circuits, so-called because each circuit is constructed from the appropriate combination of elementary gates. This stands in contrast to the topic of Chapter 5, which treats sequential circuits, or circuits with some form of memory. We begin this chapter by introducing a gate corresponding to each of the connectives discussed in Chapter 2. Next, we show how to construct a logical function from a circuit, and then move in the opposite direction. That is, how do we construct a circuit given a functional specification. This can be done in one of three ways. First, there is the direct realization in which each connective in the function is represented by a single gate. Next, the function can be realized through either its canonical SOP or POS representation, in accord with the discussion of these forms in Chapter 2. Finally, the notion of minimized circuit is introduced, also in accord with minimization technique previously discussed.

Elementary Gates

The connectives and their gate counterparts are shown in Table 3.1. The NOT gate or inverter corresponds to the logical operation of negation and has one input line. All of the other gates take two inputs. Notice that the symbols for NAND and NOR gates are identical to their nonnegated counterparts with the exception of the bubble before the output line. This bubble is the symbol for negation, and as we will see shortly, it may appear on the input line as well as the output line. In fact, the symbol for negation itself is actually composed of a buffer (the triangle), which serves to regulate the voltage level on the line (see the discussion on properties of gates below) but has no logical effect, and the inverting bubble. Table 3.1 also reprises the truth tables for each of the operators corresponding to the gates as given in Chapter 2.

TABLE 3.1 Logical Operators and the Corresponding Gate Symbol

Gate Name	Symbol	Truth Table		
NOT[1]		P1	P1'	
		0	1	
		1	0	
OR		P1	P2	P1 + P2
		0	0	0
		0	1	1
		1	0	1
		1	1	1
AND		P1	P2	P1 · P2
		0	0	0
		0	1	1
		1	0	1
		1	1	1
NAND		P1	P2	P1 NAND P2
		0	0	1
		0	1	0
		1	0	0
		1	1	0
NOR		P1	P2	P1 NOR P2
		0	0	1
		0	1	0
		1	0	0
		1	1	0
XOR		P1	P2	P1 XOR P2
		0	0	0
		0	1	1
		1	0	1
		1	1	0

NOTE

There is an alternative view of what logical gates do that will prove useful throughout the book. Rather than merely logically transforming their inputs, they can be viewed as filters that selectively allow possible dynamic input streams (i.e., a string of data composed of 0s and 1s) to pass through. For example, let us designate the first line in a two-input AND as the control signal, and the second as the data signal. The data line contains a varying succession of low and high signals. We know from Chapter 2 that $1 \cdot X \equiv X$ and that $0 \cdot X \equiv 0$. Thus, if the control signal is 1, then the output of the gate will reflect whatever happens to be on the data line at that time. If the control signal goes to 0, then the output of the AND gate will go to 0, regardless of what appears on the data line. The control signal plays the opposite role in the case of an OR gate. When it is 0, the gate will reflect the value of the data line, and when it is 1, the gate will always output 1. This follows from the fact that $0 + X \equiv X$ and $1 + X \equiv 1$. XOR provides a more complex filter. If the control signal is 0, then the bits on the data line are transmitted unaltered. However, when the line goes to 1, the bits are inverted. This follows from the fact that $0 \text{ XOR } X \equiv X$ and $1 \text{ XOR } X \equiv X'$ (see Table 3.1). Thus, XOR can provide selective inversion, with the control line indicating whether or not the data on the data line is inverted.

In the case of AND and OR and their negations NAND and NOR, it is typical to have gates that have a fan-in of greater than two, that is, with more than two input lines. For AND and OR this simply means the product or sum is extended to include the indicated number of variables. In the case of NAND and NOR the respective sum or product is formed first, and then the negation is taken. The truth table in Table 3.2 shows the corresponding function and truth values for the three-input versions of these gates. Gates with greater than three input lines are similarly formed.

TABLE 3.2 Logical Operators and the Corresponding Gate Symbol

P	Q	R	3-input OR (P + Q + R)	3-input NOR (P + Q + R)'	3-input AND (P · Q · R)	3-input NAND
0	0	0	0	1	0	1
0	0	1	1	0	0	1
0	1	0	1	0	0	1
0	1	1	1	0	0	1
1	0	0	1	0	0	1
1	0	1	1	0	0	1
1	1	0	1	0	0	1
1	1	1	1	0	1	0

Systems NOR4 and NAND4 on the CD-ROM give the truth tables for a four-input NOR gate and four-input NAND gate respectively. From Tables 3.1 and 3.2 you might conjecture that the truth values for NOR always contain a 1 in the top row and 0 elsewhere, regardless of the number of variables, and you would be correct. This is because the sum of all the variables will be 0 only when all the variables are 0, and therefore NOR, which is the complement of this sum, is 1 only when this is the case. Likewise, NAND consists of only one 0 in the final row. The reason for this is that the product of all variables is 1 when they are all 1, and therefore NAND, which is the complement of this product, is 0 only when this is the case.

Circuits to Functions and Truth Tables

The operation of a digital combinational circuit can be completely characterized in two complementary ways. First, the corresponding logical function can be derived. This is simply a matter of working through the gates one by one systematically. For example, Figure 3.1(A) shows a circuit with three gates and three total inputs, each of which will correspond to a variable in the function to be derived. It is possible to work from left to right, deriving this function from the outside in, or from right to left, deriving the function from the inside out. We will use the latter method because it is consistent with the tree-graph formalism to be developed in the next section. The steps in the derivation are as follows:

1. First, we note that the rightmost gate is an OR gate, with two input lines. Thus, we write down the corresponding connective, "+", between two expressions to be filled in, yielding the following template: (() + ()).
2. Next, we fill in the first set of parentheses, corresponding to the top NAND gate. The function with this filled in is: ((P NAND Q) + ()).
3. Finally, we turn to the right set of parentheses corresponding to the bottom input. This is simply the negation of R, and thus the final function is ((P NAND Q) + (R')), or more simply, (P NAND Q) + R'.

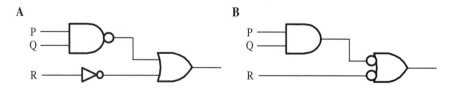

FIGURE 3.1 (A) A simple combinational circuit, and (B) an alternative representation of the same circuit.

Before turning to a more complicated example, let us look at a different representation of the same circuit in 3.1(B). Here, the bubble on the NAND gate and the inverter have transferred to the input lines on the OR gate. This is a perfectly acceptable alternative representation. It should be noted however, that in this case the bubbles are not actually separate gates, but are incorporated into the OR gate (with these inverters in place, the OR becomes a NAND, as we will shortly prove).

A more difficult example is shown in Figure 3.2. Here, we will simply list the steps in the derivation of the function; you should be able to reconstruct the reasoning behind these steps with what we've discussed so far.

1. (() XOR ())
2. ((() () ())' XOR ())
3. (((A + B) (C) (D'))' XOR ())
4. (((A + B) (C) (D'))' XOR (CD))

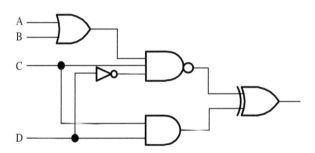

FIGURE 3.2 A more complex combinational circuit.

Once the function corresponding to the circuit has been constructed, it is trivial to derive the truth table associated with the circuit, as it is sufficient to produce the truth table corresponding to the derived function by the methods in Chapter 2. However, it is also possible to produce the truth table directly, as illustrated in Figure 3.3, which shows the same circuit in Figure 3.1(B). For this circuit, there are three input variables, so on each input line we write the $2^3 = 8$ rows of the corresponding column in the truth table. Then it is possible to propagate these strings through each gate by applying the relevant operator in a bitwise fashion. For example, applying the AND gate at the top left to 00001111 and 00110011 yields 00000011 (the new string has 1s only when the two input strings are both 1). Note that the computation of the effect of the inverters on the inputs to the OR gate on the string of values is especially easy; you just need to flip all the bits. You should confirm that the final string of values is identical to that which would have been produced had we created a truth table for the previously derived function, (P NAND Q) + R'.

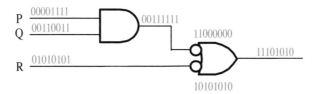

FIGURE 3.3   Deriving the truth table corresponding to a circuit.

Realizing a Function Directly

Typically, you will be asked to realize a circuit given a function rather than the reverse. For example, suppose you are given one of the functions from the last section, (((A + B)(C)(D'))' XOR (CD)), and you didn't already have the corresponding combinational circuit. One way to approach this sort of problem is first to convert the function to what is known as a tree graph, as shown in Figure 3.4. Each node in the tree corresponds to an operator, and the leaves (the bottommost nodes) correspond to variables or their negations. In order to derive the root (topmost node) of the tree, it is necessary to find the most embedded operator. This operator will be between the highest level of parentheses. For example, we can rewrite the current function of interest as [(A + B)(C)(D')]' XOR [CD], with the outer (highest-level) parentheses "[" and "]" and the lower level ones "(" and ")". XOR is between the former and therefore we put this at the top of the tree. The left side of the tree then consists of a NAND with three inputs (and therefore three branches), and the right side consists of a product with two branches. Taking the former first, the leftmost branch is itself an operator, a sum of two variables A and B, which forms the leaves. The middle branch is just C and the rightmost branch is D'. Now consider the right branch off of the original XOR. This is simply the product of C and D, and is illustrated accordingly.

The value of representing the function in tree form can be seen immediately by comparing Figure 3.4(B), which shows the tree turned on its side, and Figure 3.2, which shows the desired circuit. In order to form the circuit from this tree, you simply need to replace each operator with the corresponding gate, and then construct the input lines in accord with the leaves of the tree, which are now on the left. Note that typically we do not repeat variables. If a given variable, such as C in this case, is a leaf for more than one operator, we draw multiple lines in the circuit from the variable to the appropriate gates, with a dot indicating a branch on the line. It may also be necessary to add an inverter, as is done for the variable D in this case.

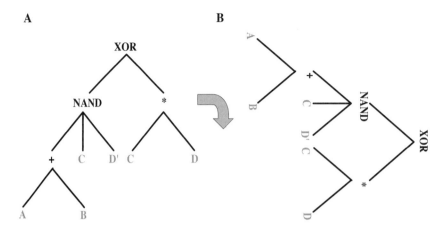

FIGURE 3.4 (A) The tree corresponding to a function, and (B) the tree turned on its side.

When constructing a function using LATTICE, it is easiest to follow the tree representation of that function. For example, to build the function illustrated in Figure 3.4 from scratch you would start by changing the "+" operator to an XOR. The left-hand side is a product of three clauses, so you would then expand the P1 on the left two times to produce this product, and change the sums to products. Then you could expand the first element in the product to produce the rightmost sum in the function. The right-hand side of the original "+" should also be expanded and turned into a product. Of course, along the way you will set the complementation of variables as appropriate. In particular, to complement the entire clause on the left-hand side of the XOR, you need to negate the leftmost "?" operator because in the tree this corresponds to the NAND operator that is above the three products on the left. When finished, you can compare your function with that already provided in the tree system, as found on the CD-ROM.

In addition to construction, LATTICE internally stores all functions as trees.

The reason for this is somewhat technical but may interest those who have some programming experience. It turns out that there is a very elegant recursive method for evaluating arbitrary logical functions once they are in tree form. One merely needs to evaluate successive branches of the tree starting from the top. For example, in our sample function, the evaluation would begin "left-side XOR right-side." The system then attempts to evaluate the left side and right side by recursively calling the original evaluation function. The algorithm proceeds this way until it eventually gets to the leaves of the trees, which are variables and therefore are 0s or 1s (the values of the variables will depend on which row in the table is under evaluation). These evaluations propagate back up the tree until an evaluation for the whole tree is produced.

Once you get used to forming circuits from functions, there is no need to go through every step in the process, and you can proceed more intuitively. Still, it is good to have a formal procedure for deriving the desired circuit in the case of complex functions.

Realizing a Circuit Through Minterms and Maxterms

Recall from Chapter 2 that any function can be transformed into a sum of products (SOP) representation, or a product of sums (POS) representation, via the minterms and maxterms of the function, respectively. Likewise, any function can be *realized* in terms of these representations. For example, let us consider the function corresponding to the circuit in Figure 3.1, (PQ)' + R'. The truth table for this function is shown in Table 3.3, with the minterms and maxterms also indicated.

TABLE 3.3 A Function and the Corresponding Minterms and Maxterms

P	Q	R	(PQ)' + R'	Minterm	Maxterm
0	0	0	1	P'Q'R'	P + Q + R
0	0	1	1	P'Q'R	P + Q + R'
0	1	0	1	P'QR'	P + Q' + R
0	1	1	1	P'QR	P + Q' + R'
1	0	0	1	PQ'R'	P' + Q + R
1	0	1	1	PQ'R	P' + Q + R'
1	1	0	1	PQR'	P' + Q' + R
1	1	1	0	PQR	P' + Q' +R'

The minterm representation of this function is the sum of minterms where the function is 1, or (P'Q'R' + P'Q'R + P'QR' + P'QR + PQ'R' + PQ'R + PQR'). The circuit realizing this function is shown in Figure 3.5(A). The circuit consists of a single OR gate, with a fan-in of seven lines, with each of these input lines corresponding to a product term in the original expression. Each of these terms is represented by an AND gate, with the inputs to a given gate either inverted or not inverted, corresponding to complemented and noncomplemented variables in the product term, respectively. For example, the inputs to the top AND gate are all inverted, corresponding to the first product term P'Q'R', and in the last term only the R line is inverted, corresponding to the last product term PQR'.

The maxterm representation of a function is the product of the maxterms when the function is 0, or in this case, just the single term (P' + Q' + R'). Thus, the realization of this function can be achieved with a single OR gate and is shown in Figure 3.5(B). If there were multiple terms in the POS representation, then we would use a single OR gate for each term and an AND gate on the right with a fan-in equal to the number of these terms.

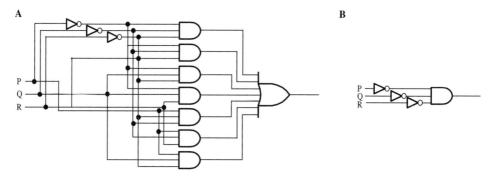

FIGURE 3.5 (A) The realization of the SOP form of $(PQ)' + R'$, and (B) the realization of the POS form.

Alternative Representations of SOP and POS Functions

An SOP function can be realized by a two-level circuit consisting of an AND gate for each product term and a final OR gate. Likewise, a POS function can be realized by a two-level circuit consisting of an OR gate for each sum term and a final AND gate. In both cases, however, there are alternative representations. In the top left of Figure 3.6(A), a hypothetical circuit for an SOP function is shown. This can be shown to be equivalent to the circuit at the bottom, consisting of only NAND gates, in two steps. In the first step (the center box), we add inverters to the end of each AND gate and to the inputs to the OR gate. It is easily seen that this does not change the effective computation of the circuit because inverting a signal twice doesn't change the signal. In the next step, the OR gate with inverted inputs is converted into a functionally equivalent NAND gate. This is possible because of a generalized form of DeMorgan's law,

$$(P_1' + P2' + P3' + \ldots Pn') \equiv (P1P2P3\ldots Pn)'. \tag{3.1}$$

In other words, the sum of complemented variables is equivalent to the complement of the product of the variables. Figure 3.6(B) shows this equivalence in gate form, for n = 3. In conclusion, any two-level circuit with AND gates in the first level and an OR gate in the second can immediately be converted into one consisting of only NAND gates.

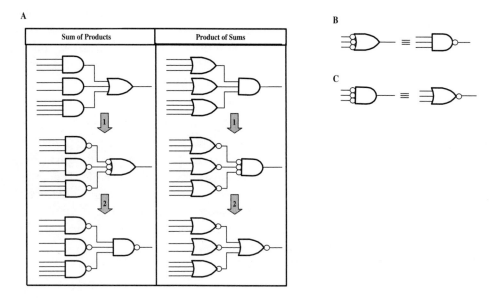

FIGURE 3.6 (A) The justification for the alternative representation of circuits realizing SOP and POS expressions in terms of NAND and NOR gates respectively, (B) the equivalence of an OR gate with all inputs inverted and a NAND gate, and (C) the equivalence of an AND gate with all inputs inverted and a NOR gate.

A similar argument can be made for the hypothetical two-level circuit shown on the right of Figure 3.6(A). In the first step, double inverters are added, and as before, the computation performed by the circuit is unchanged. The second step implicitly invokes the dual of Equation (3.1),

$$(P1'P2'P3'\ldots Pn') \equiv (P1 + P2 + P3 + \ldots Pn)'. \tag{3.2}$$

Figure 3.6(C) shows this equivalence in gate form, for n = 3. In conclusion, any two-level circuit with OR gates in the first level and an AND gate in the second can immediately be converted into one consisting of only NOR gates.

Systems DeMorgan1 and DeMorgan2 on the CD-ROM illustrate Equations 3.1 and 3.2, respectively, when the number of variables is three. Of course, in any finite truth table, it is not possible to demonstrate these general laws, which involve arbitrary numbers of variables. Those with a mathematical bent can try to prove these laws; the simplest method is probably to proceed by induction. That is, assume that the laws are true for n variables and then show that they also hold for $n+1$ variables.

Why go through the trouble of converting a circuit into a seemingly more complicated one? The answer will become apparent in the following discussion on the internal construction of gates. We will discuss why NAND gates, and to a lesser extent NOR gates, are cheaper to construct than their noninverted counterparts.

Realizing a Minimized Form of a Function

Typically we will want to implement the minimized SOP or POS description of a function rather than the original form of these expressions. The procedure mirrors the one developed in Chapter 2. First, the truth table for the given function is formed. Next, the Karnaugh map is constructed from this table. Then, the minimal number of prime implicants that cover the 1s (if the target is an SOP representation) or cover the 0s (if the target is a POS representation) is formed. If the goal is to realize a minimized form of the original function, then a circuit corresponding to one of these resulting functions is constructed.

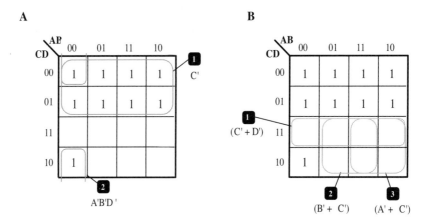

FIGURE 3.7 The Karnaugh maps and indicated prime implicants for the function [(A + B)(C)(D′)]′ XOR [CD]. (A) the PIs for the SOP minimization, and (B) the PIs for the POS minimization.

For example, let us reprise the function corresponding to the circuit in Figure 3.2, [(A + B)(C)(D')]' XOR [CD]. The SOP for this function is Σ_{ABCD}(0, 1, 2, 4, 5, 8, 9, 12, 13), as can readily be confirmed with a truth table. The Karnaugh maps for this function are shown in Figure 3.7(A) (for the SOP form), and 3.7(B) (for the POS form), and thus the minimized forms of the function are C' + A'B'D' and (C' + D')(B' + C')(A' + C') respectively. The circuit corresponding to the minimized SOP is illustrated in Figure 3.8(A). Only NAND gates are used in this circuit in accord with the derivation in the previous section (the single literal C' becomes a single noninverted input to the final NAND). Figure 3.8(B) shows the circuit containing only NOR gates for the minimized POS form. In both cases, these circuits are considerably less complex than the circuits that would be required to realize the nonminimized form of the function. The first circuit in particular has half the number of gates (not counting inverters, which are relatively cheap to implement) of the circuit representing the original function, as shown in Figure 3.2. Out of all the possible means of implementing the function, this would be the cheapest to realize.

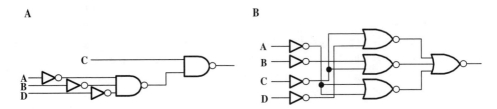

FIGURE 3.8 (A) The realization of the minimized SOP form of the function [(A + B)(C)(D')]' XOR [CD], and (B) the realization of the minimized POS form.

GATES AND INTEGRATED CIRCUITS IN PRACTICE

Logic operates outside of the domains of space and time, and so far, we have been acting as if digital circuits behave identically. But of course, this is not the case. There are a number of performance-related issues that must be understood even by the beginning circuit designer. For example, in practice, all logic gates are subject to time delays of varying degrees. If this were not true, then all computers including my old Mac Classic (the one with the 9-inch screen and 8 MHz processor) would be equivalent to the fastest supercomputer.

First, we consider how logic has been implemented over the years, from early relays to current CMOS circuits. Then we discuss the correspondence between logical values and voltage levels. Next, the topic of fan-in and fan-out is treated, particularly as it affects CMOS design. Then we show how to compute the propagation delay for gates and the circuits that are built from these elements. We conclude with a brief treatment of how to implement gates from transistors and the implications of this for higher-level design. None of these treatments is comprehensive (each could easily consume a book in itself), rather, they are intended to provide a feel for the kinds of additional extra-logical issues that influence digital design.

Logic Technologies and Logic Families

Table 3.4 gives a brief history of the development of technologies that have been used over the years to implement logic. Engineers constructed the first gates from electromagnetic relays, which were set to close and therefore conduct current under varying input conditions. Relays were both slow and large, and were eventually replaced by vacuum tubes in the 1940s and 1950s. Tubes were faster but still clunky and very power hungry. The office where the author works was sometimes plunged into darkness during these years by one of the first computers, the ENIAC, which consisted of 18,000 tubes and was located four blocks away at the University of Pennsylvania. Tubes have recently undergone a minor renaissance in the high-end audio world because of their analog characteristics, but have not been used digitally in over half a century. They have been replaced by integrated circuits (ICs) or chips, first by the TTL family in the 1960s, and more recently by the CMOS family. ICs are simply collections of many transistors that combine to serve a given function or multiple functions. For example, a small-scale integrated (SSI) circuit could contain eight NAND gates and a very large-scale integrated (VLSI) circuit could contain an entire microprocessor on a chip.

TABLE 3.4 A Brief History of Logic Technologies

Technology	Date
Relays	1930s
Vacuum Tubes	1940s–1950s
TTL ICs	1960s–1990s
CMOS ICs	1990s–present

TTL (transistor-transistor logic) is based on the bipolar junction transistor. The CMOS (complementary metal-oxide semiconductor) logic family is based on the MOSFET (metal-oxide semiconductor field-effect transistor). Initially difficult to fabricate, CMOS chips have largely eclipsed TTL-based technologies because of their lower power consumption and comparable speed characteristics. For this reason, we will concentrate on this family when it's pertinent in the following sections.

Values and Voltages

In order to be useful, digital gates must somehow implement the logical values 0 and 1. This is typically done by equating the value 0 with a low voltage and the value 1 with a high voltage (it is also possible to reverse these associations). For example, an AND gate will have a high voltage on its output line, relative to ground, if and only if its two input lines are also at a high voltage. When one or both of these lines drops to low voltage, then the output will also drop to low voltage.

If it were not for the problem of noise, that would be all we could say about this topic at an elementary level. The origins of noise are numerous and include cosmic rays, external magnetic fields, and the heat generated by the circuit itself. From our point of view, we can ignore the source of the noise and treat it as a random positive or negative value that may be added at any time to any line. For example, a gate may be expecting a high voltage of 5V on a given line, but because of noise on the line, it actually receives 4.4V. For this reason, gates are designed to operate with some margin of error.

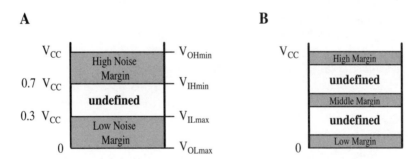

FIGURE 3.9 (A) Typical noise margins for a CMOS gate, and (B) hypothetical noise margins for a ternary gate.

Figure 3.9(A) shows the characteristic margins for a typical CMOS gate. Vcc is typically 5V for CMOS gates and the other levels are defined in terms of this upper value. V_{OHmin} is the minimum voltage at the *output* of the gate and V_{IHmin} is the minimum *input* voltage that is recognized as high for the next gate in the series. Notice that the former is set higher than the latter. Thus, if a noise disturbance lowers the output voltage of one gate that is generating a high signal, as long as the disturbance is sufficiently small, the resulting voltage will still be above the required value for a subsequent gate to see it as high. Likewise, V_{OLmax} is the maximum voltage at the output of the gate for a low signal and V_{IHmax} is the maximum input voltage that is recognized as low. Thus, a noise disturbance that raises the effective output voltage of a gate still has a chance to be recognized as a low signal as long as this disturbance is below the noise margin. The undefined region acts as a buffer between the effective low and high logic areas, with signals in this area interpreted as neither high nor low.

We are now in a position to address a problem that we originally skirted, namely, the ideal base for computers and the circuits on which they are based. We have already seen that base 12 is probably ideal for humans. As it turns out, by a measure that is believed to be critical for determining hardware complexity, a base of e (the base of natural logarithms) is best for computers (because we cannot work with noninteger bases, we will use the nearest integer, namely 3). The measure is simply the product of the number of digits it takes to represent a number with the number of symbols per digit. For example, consider representing the numbers 0 through 999,999 in decimal. This takes up to six digits, with 10 possibilities per digit, with a complexity measure of 6*10 = 60. In binary, a 20-bit number is sufficient, so the measure is 20*2 = 40. However, in base 3 (ternary), 13 digits are sufficient, so the measure is 13*3 = 39, which is slightly better than base 2.

One question remains: is the small gain in moving from base 2 to base 3 worth it from a hardware point of view? The answer is probably not. First, there is the problem of designing ternary gates that operate efficiently from the point of view of speed and power, which is no trivial matter in itself. Figure 3.8(B) shows another problem. If the undefined buffer regions are kept the same as in 3.8(A), then out of necessity, the noise margins (three in this case for each of the three states) must be reduced. Thus, a ternary circuit will be more sensitive to noise disturbances. Finally, there is the matter of retooling thousands of electrical engineers to think and design in ternary. The interested reader can turn to http://www.americanscientist.org/template/AssetDetail/assetid/14405/page/3 for an excellent discussion of these issues.

Fan-In and Fan-Out

Logical expression contains no fixed boundaries in the sense of the number of variables an operator can take or the degree of embedding within the expression. However, the digital realizations of functions are subject to a number of extra constraints. Two of the most crucial are fan-in and fan-out; that is, the number of input lines into a gate and the number of effective outputs from a given gate to others. For example, the fan-in of the NAND gate in Figure 3.2 is three and the fan-out is one. If the input line D in this diagram was actually the output of a gate, then this gate would have a fan-out of two (it loads both the inverter and the AND gate).

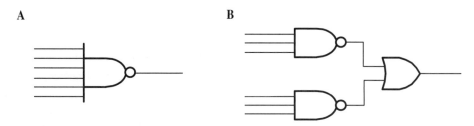

FIGURE 3.10 (A) A six-input NAND, and (B) a reformulation with two NAND gates and an OR gate.

Fan-in for CMOS gates is usually highly constricted, to about four to six lines, depending on the type of gate. Thus, it may be necessary to reformulate a circuit that would otherwise have a high fan-in to meet these constraints. Figure 3.10(A) shows a NAND gate with a fan-in of six. Let us assume that maximal allowable fan-in is four. Figure 3.10(B) shows one way of circumventing this constraint. The NAND gate has been replaced with two gates, each with a fan-in of three, and these then feed into an OR gate. The two circuits are equivalent. Why? (ABC)' + (DEF)' ≡ (A' + B' + C' + D' + E' + F') ≡ (ABCDEF)', by two applications of DeMorgan's law.

ON THE CD

System reformulate on the CD-ROM shows the equivalence of (P1P2)' + (P3P4)' ≡ (P1P2P3P4)'. Screen width limitations prevent LATTICE from illustrating a longer version of this law. It is not difficult to show mathematically however, that it is true for an arbitrary number of variables, that is, (P1P2...Pm) ' + (P(m +1)P(m + 2)...Pn) ' ≡ (P1P2...Pm...Pn) '. There is an even more general form in which the complement of a product can be divided into an arbitrary number of sums of complements of the subsets comprising the set of variables in the original product (see Exercise 3.6).

The fan-out situation for CMOS gates is a bit more complex. Propagation delay (the time for a gate to respond) typically increases as fan-out increases, usually linearly. This means that after a certain point, increasing fan-out will cause a gate to exceed its listed maximum delay (the next section contains a more detailed discussion of this and other delay parameters), possibly leading to undesirable behavior. Increasing fan-out beyond this point eventually will lead to a disruption of the noise margins, making the gate unusable. In summary, there are both soft (possibly violable) and hard constraints (inviolable) that must be considered in the case of fan-out. In both cases, it may be necessary to reformulate the circuit using similar techniques as shown in Figure 3.10.

Gate Delays and Circuit Delays

Here we consider two aspects of timing in digital circuits: the delay associated with an individual gate and the delay associated with a circuit as whole. As previously mentioned, a gate will be subject to a delay if its output level switches (if this does not occur, then there is no effective delay even if the input lines change because the output remains at a constant level). Figure 3.11 illustrates a typical delay for a gate, in this case an inverter. The figure shows the time course of the voltage levels when the input voltage goes from low to high and back to low. The output voltage follows this transition in the opposing directions, in accord with the inverter's function. The other salient features of the diagram are as follows:

- There is a rise or fall time associated with a change in voltage levels. This can be seen in the diagonal transitions; if there was no rise or fall time, these ramps would be vertical.
- Output voltage changes lag behind input levels. These propagation delays t_p are measured from the center of the input ramp to the center of the output ramp.
- The high to low transition delay in the output voltage, t_{pHL}, is generally not equal to the low-to-high delay, t_{pLH}. In this case, the former is shorter than the latter.

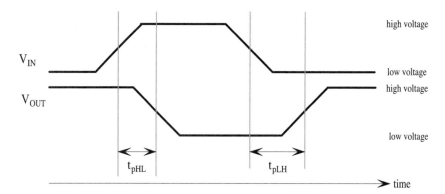

FIGURE 3.11 The time course of voltage levels for both the input and output when a low voltage followed by a high voltage followed by a return to low voltage is applied to the input of an inverter.

Different gates will have different propagation delays, and in the case of CMOS gates, this will typically be a function of the fan-out of the gate in addition to the nature of the gate and whether the output signal rises or falls. Determining the typical and maximum delays of a circuit is a complex matter and is usually accomplished with the aid of a simulation program. Nevertheless, we can provide a feel for the kinds of calculations involved with the following example.

Suppose the delays for differing gates are as shown in Table 3.5 (these numbers are hypothetical, but are similar to those for fast CMOS gates). Two sets of values correspond to each gate, typical delays and maximum. Typical values represent average delays and maximum values represent worst case scenarios; gates are guaranteed not to exceed these latter values. High-to-low and low-to-high delays are given for each set.

TABLE 3.5 Hypothetical Delays for Differing CMOS Gates in Nanoseconds (ns)

GATE	Typical		Maximum	
	t_{pLH}	t_{pLH}	t_{pLH}	t_{pLH}
Inverter	5.5	6.0	9	9.5
OR	5.0	5.0	8.5	8.5
AND	5.0	5.5	8.5	9.0
NAND	4.0	5.0	9.0	9.5
XOR	5.0	5.5	10.0	11.0

Let us assume that in one of our characteristic circuits, Figure 3.2, all gates initially have a low output and that line A is set to high, B to low, C to low, and D to high. First we will calculate the expected typical delay for this input combination. The typical delay on the top OR gate is then 5.0 ns and the delay on the inverter is 6.0 ns (it goes from high to low). Thus, we can assume that the top NAND begins its work when the last input arrives, or at 6.0 ns, and it adds 4.0 ns to this (it goes from low to high). The total so far is 10.0 ns. We can ignore the bottom AND gate because it does not change state, and move to the final XOR gate, which adds 5.5 to the total when it goes from low to high output. Thus, the total typical delay time for these sets of inputs is 15.5 ns.

One use of the data in Table 3.5 is the construction of an absolute worst case scenario. This is done by taking the maximum of the maximum delay calculations over all possible input sets. This number is useful because we know that the circuit will be guaranteed to compute faster than this value under all possible circumstances. In order to perform this calculation, we need to consider the maximum delay associated with every possible input combination, as illustrated in Table 3.6. Let us consider for example, what happens when the inputs are 0001, corresponding to the second row of the table. Level 1 of the circuit comprises the OR gate at the top of the circuit and the inverter. The OR gate does not change state, but the inverter goes from low to high. Consulting Table 3.5 we see that the delay for this change is 9.5 ns. Level 2 comprises the NAND gate and the AND gate at the bottom of the circuit. The NAND gate goes from low to high with this set of inputs, resulting in an additional maximal delay of 9 ns. The AND gate does not change state. Finally, level 3 is just the XOR gate; in this case, it goes from low to high for an additional 10 ns delay. Thus, the total maximal delay for this input combination is 28.5 ns.

Another example may help to illustrate this process. When the input is 0110, the inverter goes from high to low, but all other gates remain in their original low states. In particular, the XOR does not change state, which means that there is no delay whatsoever. This is somewhat counterintuitive—it seems that the XOR gate must at least wait for the other gates to perform their calculations before it decides what its output should be. However, with this set of inputs, the wait is pointless. The XOR gate starts low and stays low; thus the output is correct from the start and there is no delay.

The other cases are computed accordingly. The worst-case delay will be the maximum of these maximal delays, or 28.5 ns. This occurs four times for this circuit: when the inputs are 0001, 0101, 1001, and 1101 (these cases are shaded). If this circuit is feeding into another circuit, we know that under the worst circumstances, when the gates are particularly bad and the input combination is just right, then the downstream circuit may have to wait as much as 28.5 ns before it can begin its computation.

Implementation of Gates

The astute reader may wonder why we went to the trouble of implementing SOP expressions with only NAND gates instead of the more natural combination of AND gates for the first level and a single OR gate in the second. This section describes the reason. As it turns out, generally it requires fewer transistors to implement inverting gates (such as a NAND and NOR gate) than noninverting gates (such as AND and OR gates). For example, Figure 3.12(A) shows a switch model for a CMOS NAND gate and a CMOS AND gate is shown in Figure 3.12(B). The former requires two fewer transistors then the latter.

TABLE 3.6 Maximum Delay for Each Input Combination for the Circuit in Figure 3.2.

ABCD	Level 1	Level 2	Level 3	Total
0000	9	9	10	28
0001	9.5	9	10	28.5
0010	9	9	10	28
0011	9.5	9	0	18.5
0100	9	9	10	28
0101	9.5	9	10	28.5
0110	9	0	0	0
0111	9.5	9	0	0
1000	9	9	10	28
1001	9.5	9	10	28.5
1010	9	0	0	0
1011	9.5	9	0	0
1100	9	9	10	28
1101	9.5	9	10	28.5
1110	9	0	0	0
1111	9.5	9	0	0

The way these models work is as follows. Two kinds of transistors are used: p-channel and n-channel MOSFET. For our purposes, it suffices to note that a p-channel transistor will act as a closed switch if the input is low, and an n-channel transistor will act as a closed switch if the input is high. For example, the two inputs, both low for the purposes of illustration, close the two p-channel switches at the top and open the two n-channel switches at the bottom of Figure 3.12(A). The net effect of these actions is to close the circuit from the input voltage V_{DD} to the out, as desired (two low inputs to a NAND gate generates a high output). The only time that the output line is both disconnected from V_{DD} *and* connected to ground (the bottom triangle) occurs when both inputs are high, in accord with the fact that this is the only case when NAND produces a low output.

The simplest way to implement an AND gate with MOSFET transistors is shown in Figure 3.12(B). This is just the previous circuit with the addition of two transistors that realize an inverter. In the case shown where both inputs are low, the output line is connected to ground. The only case in which this line is connected to the input voltage V_{DD} and disconnected from ground occurs when both inputs are high.

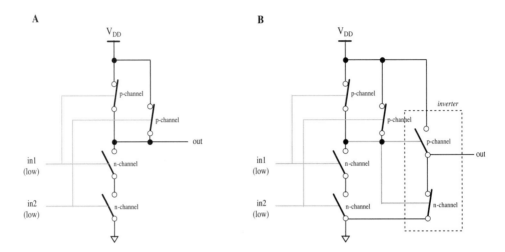

FIGURE 3.12 (A) A switch model for a NAND gate, and (B) the same model for an AND gate, which has an added inverter.

ON THE CD

System NANDimpl on the CD-ROM illustrates how the switch model implements a NAND gate. Two possible connections can be made, labeled conn1 and conn2. The first connects V_{dd} and out, and the second connects out and ground. Clicking on conn1 shows its truth function, input1' + input2'. This reflects the fact that when either input is low, then one of the parallel p-channels will be closed, and thus the connection will be made (see Figure 3.12). Clicking on conn2 shows its truth function, input1 Σ input2. Both of these n-channels must be closed for the connection from out to ground to be made. The net effect of these truth functions is that out short-circuits to ground only when both inputs are high; in all other cases, V_{dd} short-circuits to out. Confirm this result by working through the four possible combinations of inputs in the input machine. System ANDimpl demonstrates the same type of model for the AND gate. In this case, the key functions are just the complements of that for NAND, reflecting the extra inverter in the model (see Figure 3.12(B)). Confirm that the only case in which V_{dd} is connected to out is when both inputs are high, or in other words, when this implements an AND gate.

The switch model demonstrates, surprisingly, that noninverting gates are implemented by adding an inverter to an inverting gate rather than vice versa. This justifies the logical steps we went through in order to show the equivalence of SOP and NAND formulations of functions. It is also true that NOR gates are internally simpler than OR gates, and therefore, should be preferred when realizing POS functions. Figure 3.13 shows one possible switch model for the NOR gate. It works similarly to the NAND gate (see Figure 3.12(A)), except that the p-channels are now in serial and the n-channels are in parallel. When both inputs are low (the case illustrated in the figure), both of the p-channels are closed and current flows from V_{dd} to out. The n-channels are open, preventing a connection from out to ground. In all other cases, there is no connection between V_{dd} and out, but at least one of the n-channels will be closed, and therefore, out will be connected to ground.

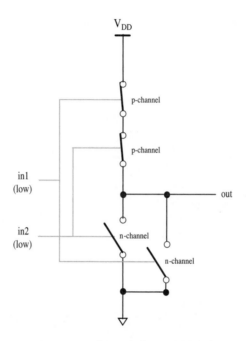

FIGURE 3.13 The switch model for the NOR gate.

Comparing Figures 3.12(A) and 3.13, we see that the NOR gate implementation requires the same number of transistors (four) as the NAND gate implementation.

However, NOR gates are typically somewhat slower than NAND gates, because p-channel transistors in series are slower than n-channel resistors in series. The former is present in the NOR gate and the latter in the NAND gate; therefore, the NAND gate is faster. From the point of view of high-level design, this implies that all things being equal, SOP formulations and their NAND implementations should be favored to POS formulations and their NOR implementations.

SUMMARY

This chapter has made the first step in the transition from logic as an abstract mathematical set of rules to the implementation of logic with gates. First, we introduced elementary gates and then we showed how to produce either the functional equivalent or a truth table corresponding to a circuit composed from a collection of such gates. The more interesting direction starts with a function and attempts to build a circuit that will realize this function. This can be done directly, or as is more efficient, from the minimized form of the function. We also showed how standard SOP and POS functions, whether minimized or not, can be realized with only inverters and NAND gates, or with inverters and NOR gates, respectively.

The second half of the chapter discussed implementation in practice, as opposed to merely mirroring the Boolean algebra techniques introduced in Chapter 2. We introduced logic families and gave a short history of digital circuits. We discussed the notion of converting logical signals into voltages and the problems associated with doing so. We raised two important considerations in practical design: limitations in fan-in and fan-out, particularly as it relates to CMOS design. Next, we treated the notion of gate delay and circuit delay, and we showed how to compute the worst-case delay for a given circuit. The chapter concluded with a brief treatment of the transistor implementation of gates, and in particular, it showed why NAND gates, and to a lesser extent NOR gates, are preferred to their noninverted counterparts.

EXERCISES

3.1 For each of the following functions:
- show the equivalent logic tree
- realize it directly with logic gates
- realize it by converting it to a minterm representation with inverters and AND and OR gates
- realize it after SOP minimization with inverters and AND and OR gates
- realize it after SOP minimization with inverters and NAND gates

(a) PR + RQ' + P'QR
(b) [(A + B + C)(D)] NAND [B'D]
(c) (A + (B NAND C') + D) XOR (B' + (A(B XOR C)))

3.2 For each of the functions in Exercise 3.1:
- realize it by converting it to a maxterm representation with inverters and AND and OR gates
- realize it after POS minimization with inverters and AND and OR gates
- realize it after POS minimization with inverters and NOR gates

3.3 For each of the following functions, show the realizations corresponding to the minimized SOP and POS expressions, and state which would be cheaper to implement, assuming that NAND and NOR gates are equivalent and inverters are free.
   (a) $\Sigma_{WXYZ}(1, 5, 6)$
   (b) $\Sigma_{WXYZ}(0, 3, 8, 10)$
   (c) $\Sigma_{WXYZ}(0, 1, 4, 5, 6, 7, 8, 9, 11, 12, 13, 14, 15)$

3.4 Table 3.7 shows the truth table for the XNOR gate (which realizes the equivalence operator). Realize this function with only NAND gates, and then with only NOR gates. (Hint: First figure out how to build an inverter from a NAND gate, then from a NOR gate, and then realize the sum of the minterms and product of the maxterms for this function.)

TABLE 3.7 The Truth Table for XNOR

P1	P2	P1 XNOR P2
0	0	1
0	1	0
1	0	0
1	1	1

3.5 What are the two gates in Figure 3.14 equivalent to?

FIGURE 3.14 The gates for Exercise 3.5.

3.6 Using the computing complexity measure (the product of the number of digits it takes to represent a number with the number of symbols per digit), determine which base most efficiently represents the numbers to 10 million and the numbers to 1 billion.

3.7 Suppose you are asked to implement a NOR gate with six inputs, but there is a fan-in limitation of four lines for the NOR gate. Design a circuit that circumvents this problem.

3.8 Show that NAND over n variables can be divided into an arbitrary number of sums of NANDS in which the union of the sets in sums is the same as the original set (for example: (P1P2P3P4P5P6P7)' ≡ (P1P2)' + (P3P45)' + (P6P7)').

3.9 Calculate the voltages for VIHmin and VILmax in Figure 3.9(A) given that Vcc = 5.0V. Calculate the significant values for 3.9(B) given the same value of Vcc.

3.10 Calculate the worst-case delay in the circuit corresponding to the function (A + B) XOR (C + D') using the values given in Table 3.5.

3.11 Let each switch in Figure 3.12 be a variable in a truth table (indicate whether the switch is open or closed depending on the input). Use this table to show that the first circuit correctly realizes a NAND gate and the second correctly realizes an AND gate.

3.12 Form an OR gate from the switch model given in Figure 3.13 in an analogous way that an AND gate was formed from a NAND gate in Figure 3.12.

LATTICE Exercises

3.13 Use LATTICE to show the equivalence of (P1P2) + (P3P4) and (P1 NAND P2) NAND (P3 NAND P4). That is, show that the transformation on the left side of Figure 3.6 is valid.

3.14 Use LATTICE to show the equivalence of (P1+ P2) (P3 + P4) and (P1 NOR P2) NOR (P3 NOR P4). That is, show that the transformation on the right side of Figure 3.6 is valid.

3.15 Modify the ANDimpl system to emulate the switch model for the OR gate given in Figure 3.13.

ENDNOTES

1. Usually we will use a smaller, but otherwise identical version of the NOT gate symbol in the circuit diagrams in this and later chapters.

4 Complex Combinational Circuits

In This Chapter

- Introduction
- Binary Adders
- Decoders and Encoders
- Multiplexers and Demultiplexers
- Programmable Logic Devices (PLDs)
- Summary
- Exercises

INTRODUCTION

As with any design process, the larger the building block, the swifter the implementation, and digital design is no exception to this rule. Rather than treating logic gates as the elements of design, it is often useful to begin with larger circuits that are in turn constructed from these gates. In this chapter, we will treat four such circuits. We begin with a treatment of binary addition, and then move to decoders and encoders. We follow this with a discussion of multiplexers and demultiplexers, and conclude with programmable logic devices (PLDs).

BINARY ADDERS

Recall from Chapter 1 that the addition of two binary numbers proceeds column by column, with the rules for each column indicated in Table 4.1. Each column i in the addition takes three inputs (two for the addends, A_i and B_i, and one for the carry bit, C_i) and produces a sum for the column S_i and a carry for the next column C_{i+1}. The outputs in this table are the result of an arithmetic computation, but there is no good reason why we should not consider them to be the output of a yet undetermined logical function. This insight, that a computational table can also be viewed as a truth table, is what will allow us to construct arithmetic circuits with digital logic in the following sections.

TABLE 4.1 The Sum and Carry for the Next Column As a Function of the Values in the Current Column for Binary Addition

Inputs			Outputs	
C_i	A_i	B_i	S_i	C_{i+1}
0	0	0	0	0
0	0	1	1	0
0	1	0	1	0
0	1	1	0	1
1	0	0	1	0
1	0	1	0	1
1	1	0	0	1
1	1	1	1	1

Full Adder

The first step in the construction of a digital adder is the construction of a circuit that is capable of producing the sum and the carry for a single column of the summation. In order to realize this circuit, first we will minimize Table 4.1 with the techniques developed in Chapter 2. Figure 4.1 shows the Karnaugh maps for the sum and carry functions.

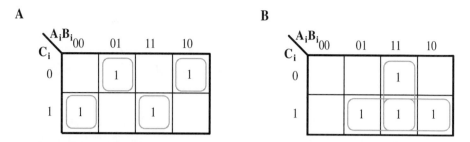

FIGURE 4.1 (A) The Karnaugh map for the sum, and (B) the map for the carry of the next column, with the prime implicants circled in each case.

The minimized SOP expressions corresponding to these maps are

$$S_i = C_i'A_iB_i' + C_i'A_i'B_i + C_iA_i'B_i' + C_iA_iB_i \qquad (4.1)$$

and

$$C_{i+1} = C_iB_i + A_iB_i + C_iA_i. \qquad (4.2)$$

These expressions can be rewritten as

$$S_i = (C_i \text{ XOR } (A_i \text{ XOR } B_i)) \qquad (4.1')$$

and

$$C_{i+1} = A_iB_i + C_i(A_i \text{ XOR } B_i). \qquad (4.2')$$

The equivalence of Equations 4.1 and 4.1' is shown by the following steps:

Step 1. $(C_i \text{ XOR } (A_i \text{ XOR } B_i)) \equiv C_i \text{ XOR } (A_iB_i' + CA_i'B_i)$
Step 2. $\equiv C_i(A_iB_i' + A_i'B_i)' + C_i'(A_iB_i' + A_i'B_i)$
Step 3. $\equiv C_i(A_i'B_i' + A_iB_i) + C_i'(A_iB_i' + A_i'B_i)$
Step 4. $\equiv C_i'A_iB_i' + C_i'A_i'B_i + C_iA_i'B_i' + C_iA_iB_i$

The first two steps in this derivation follow from the definition of XOR, the third step follows from the fact that equivalence is the complement of XOR, and the fourth step follows from the factoring rule for expressions.

The equivalence of Equations 4.2 and 4.2' is shown by the following steps:

Step 1. $A_iB_i + C_i(A_i \text{ XOR } B_i) \equiv A_iB_i + C_i(A_i + B_i)(A_iB_i)'$
Step 2. $\equiv A_iB_i + (C_iA_i + C_iB_i)(A_iB_i)'$
Step 3. $\equiv A_iB_i + C_iA_i + C_iB_i$

The first step follows from one definition of XOR: X XOR Y is equivalent to X + Y and not both X and Y. The second step follows from factoring, and the third step from the general law $(X + YX') \equiv (X + Y)$ which can be readily proved with a truth table (here X is A_iB_i and Y is $(C_iA_i + C_iB_i)$).

These equivalences can also be confirmed by enumerating the truth values of these functions (Exercise 4.1 asks you to confirm with a truth table that the expression in Equation 4.1 is equivalent to that in Equation 4.1', and that Equation 4.2 is equivalent to Equation 4.2'). The reasons for rewriting these expressions are twofold. First, Equation 4.1' is more compact than Equation 4.1 and therefore cheaper to realize. Second, Equation 4.2' contains the same subexpression, (A_i XOR B_i) as Equation 4.1', leading to greater economy when simultaneously implementing both of these functions.

This can be seen in the circuit in Figure 4.2. The top left XOR gate feeds into the right XOR gate to form the circuit that produces the sum S_i, and also feeds into the bottom right AND gate to form the product which is the rightmost term in (4.2'). The result is what is known as a full adder, so-called because it is composed of two identical half-adders (the subcircuits in the shaded boxes). This circuit embodies an important lesson in design: To achieve efficiency in a multipurpose design, strive to create economy by using gates to achieve multiple effects. In this case, this is achieved by the dual uses of the first XOR gate.

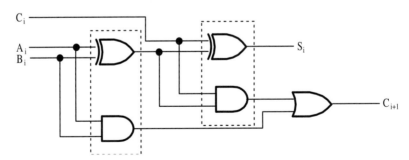

FIGURE 4.2 The full adder, composed of two half-adders (dotted boxes).

Ripple-Carry Adder

At this point, we have a circuit that is capable of adding a single column in a binary sum, but what we really want is one that is capable of adding every column in the problem. Looking at the 4-bit example below, we can see what needs to be done.

$$
\begin{array}{rr}
carry & 0110 \\
 & 0011 \\
+ & 1001 \\
\hline
 & 1100
\end{array}
$$

The carry generated by a given column needs to be routed into the input for the next column. For example, in the above case, the rightmost column generates a carry of 1. This is written at the top of the second column in the sequence, and is one of the inputs for this column.

This idea is the inspiration for the ripple-carry adder shown in Figure 4.3. Each full-adder (FA) corresponds to the circuit in Figure 4.2. This circuit computes the sum for a given column and then generates two items. The first is the sum as indicated by the line leaving from the bottom of the FA. The second is the carry, leaving from the right of the FA, which is then routed into the FA responsible for computing the sum of the next bit. In other words, the ripple-carry adder is the exact analog of the process that occurs when manually computing a binary sum; the carry generated by the ith column is forwarded to the i+1st column.

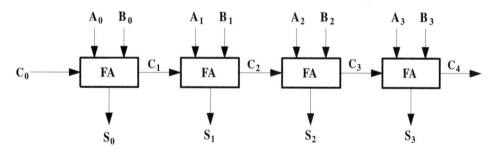

FIGURE 4.3 The ripple-carry adder.

The ripple-carry adder is a good example of an iterative circuit. The defining characteristic of such circuits is the repetition of a key component, with one component feeding into the next in order to achieve design economy. Compare the operation of the ripple-carry adder with a circuit realizing the truth table for the binary addition problem directly. This table, for a 4-bit addition problem, would consist of

$2^8 = 256$ rows corresponding to each of the possible eight input variables (four for each addend), with four outputs, one for each of the sums. Even after minimization, this circuit would contain a far greater number of gates than the corresponding ripple-carry adder. Some sort of iteration is even more crucial in the arithmetic logic unit (ALU) of modern computers, which typically process 4- or 8-byte (32 or 64 bit) numbers.

ON THE CD

System ripple in the Chapter4 folder on the CD-ROM is identical to the addition system we first saw in Chapter 1. Now however, we are in a better position to understand how it works. Clicking on any cell in the sum machine shows that the truth table for these cells is that given in Equation 4.1'. Clicking on "var" reveals that they are responding to the presence of the carry cell and the two addend cells respectively. Clicking on any cell in the carry machine shows that the truth table for these cells is that given in Equation 4.2'. Clicking on "var" reveals that the next state of each carry cell is a function of the carry and the addends in the previous column, in accord with Equation 4.2'. This system works in the same way that the ripple-carry adder does, by propagating carries from right to left until the computation is complete.

Carry–Look-Ahead Adder

Iterative circuits achieve design economy but at the expense of processing speed. For example, the ripple-carry adder needs as many stages as there are bits to add, and each stage must wait for the one before it to finish before generating its response. If, however, there was some way of generating the carries without having to go through the intermediate processing, then later stages could proceed without this delay. This is the idea behind the carry–look-ahead adder.

The derivation of a general expression for the carry signal begins with the following observation. A carry in the i+1 column is generated in two cases: (1) both of the addends in the ith column are 1 or (2) one of the addends is 1 and the carry for the ith column is 1 (refer to Table 4.1 for confirmation of this fact). In Boolean algebraic terms, we have the following, reflecting these two conditions respectively:

$$C_{i+1} = A_i B_i + (A_i \text{ XOR } B_i) C_i. \qquad (4.3)$$

We can rewrite this as

$$C_{i+1} = G_i + P_i C_i, \qquad (4.3')$$

where $G_i = A_i B_i$ and $P_i = (A_i \text{ XOR } B_i)$. G_i is called the generator because it generates a new carry when both A_i and B_i are 1, and P_i is called the propagator because it propagates the previous carry whenever A_i or B_i is 1 (if both are 0 or both are 1, then P_i is 0 and no propagation occurs).

Equation 4.3' is known as a recursion relation, and can be used to generate the carry for any column as a function solely of the addends in the prior columns. To demonstrate how this works, let us calculate C_{i+2}.

$$\begin{aligned} C_{i+2} &= G_{i+1} + P_{i+1}C_{i+1} \\ &= G_{i+1} + P_{i+1}(G_i + P_iC_i) \\ &= G_{i+1} + P_{i+1}G_i + P_{i+1}P_iC_i \end{aligned} \quad (4.4)$$

This derivation proceeds in two steps. First, C_{i+2} is expressed in terms of the G_{i+1}, P_{i+1}, and C_{i+1} in accord with Equation 4.3' (this expression is true for all i, and thus holds between any column and its predecessor). In the second step, C_{i+1} is expressed in terms of G_i, P_i, and C_i, also following Equation 4.3'. The net effect of these two steps is to bypass C_{i+1} and express C_{i+2} in terms of the addends of the previous columns (which are available at the start of the computation), and C_i. Thus, this carry can be generated without waiting for the carry immediately preceding it to be generated. Equation 4.3' can be repeatedly applied to generate the carry without having to wait for the previous n carries, although as n grows, so does the resulting expression and the corresponding hardware (see Exercise 4.3 and the following discussion).

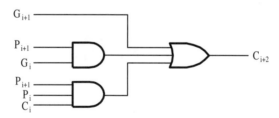

FIGURE 4.4 The realization of a two-stage carry–look-ahead signal.

Implementation of the carry–look-ahead could not be simpler. We already know how to create the G_i and P_is; these correspond to the outputs of the first half-adder in Figure 4.2. Once we have these signals, Equation 4.4, which describes a look-ahead of two, can be realized with the two-level SOP implementation shown in Figure 4.4. Carry–look-ahead circuits such as this will always contain just two levels regardless of the degree of look-ahead, allowing efficient computation of the carry in all cases, although the number of AND gates will grow as the look-ahead increases. At some point, so many gates are needed to compute the look-ahead that the circuit becomes unwieldy. Thus, there is a trade-off between the amount of hardware dedicated to the look-ahead process and the cost of implementation, with the balance between these competing ends determined by the circuit designer.

Space-time tradeoffs are a recurring theme in both computer hardware and software. If there were virtually an infinite amount of storage space, then one could enumerate the behaviors one would like to achieve in a look-up table and achieve excellent performance. Alternatively, if there were virtually no delay associated with gates, then one would always use an iterative method regardless of the number of levels of computation. In real life and for real problems, the best design lies somewhere between these extremes. For example, in this case, you would need to compute chip size and chip cost as a function of the number of degrees of look-ahead. You would also compute speed increase as a function of this same variable. At that point, you would decide the appropriate degree of look-ahead that conforms to the desired budget and achieves the required degree of performance.

System look-ahead on the CD-ROM contains a simple but representative adder that performs some advance calculations. In particular, the look-ahead cell is able, in a single step, to compute the carry value for this column. The function for this expression can be computed from recursion relation Equation 4.3' or Equation 4.4. However, with a look-ahead of only two, it is simpler to consider all the possibilities for the four addend cells to the left. There are six instances (see Exercise 4.4) in which the carry should be 1 as a function of these four variables, and after minimization the function becomes $A_2B_2 + A_1B_1B_2 + A_1B_1A_2$, as can be seen in the Karnaugh map in Figure 4.5. With this look-ahead in place, and the other cells as in a normal carry, this system becomes a more efficient adder than the prior ripple system. For example, when adding 011 to 001, it would ordinarily take two time steps for the look-ahead cell to turn one, and then one more for this to affect the sum. In this case, it takes only one time step, and the sum is computed in a total of two time steps.

Two's Complement Addition and Subtraction

As you may recall from Chapter 1, the purpose of representing numbers in two's complement is to make binary addition more efficient. Once numbers have been converted to their two's complement form, they can be added as ordinary binary numbers, regardless of whether they are positive or negative. Thus, the ripple-carry adder in Figure 4.3, or one modified with a carry–look-ahead, is perfectly adequate for this task. There are two modifications we may wish to make, however. The first performs conversion of a positive number to a negative, and the second detects overflow.

Complex Combinational Circuits

A_2B_2 \ A_1B_1	00	01	11	10
00				
01			1	
11	1	1	1	1
10			1	

FIGURE 4.5 The Karnaugh map for the look-ahead cell.

In order to subtract one two's complement number from another, the second number must be negated. Recall that the method for doing so is to flip all the bits and then add 1 to this number. We could build a special circuit to do this, which we would invoke when subtraction rather than addition was called for, but Figure 4.6 shows a more elegant solution in the context of the ripple-carry adder. The M line is on only when subtraction is indicated. When this is the case, the B_is are inverted (Why? 1 XOR 0 = 1, and 1 XOR 1 = 0). Therefore, the XOR gates carry out the bit flipping on B when subtracting is desired. There still remains the need to add 1 to the result. However, this can be accomplished (cleverly) with the routing of the M line into the initial carry C_0, which is ordinarily 0. This adds 1 to the sum, which is the same as adding 1 to B: $(A + (B + 1)) = (1 + (A + B))$.

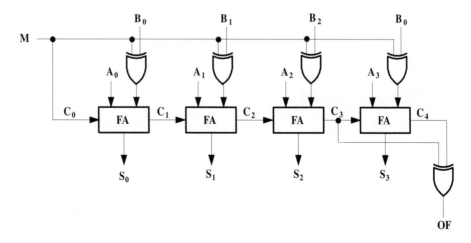

FIGURE 4.6 The realization of a two's complement adder with overflow detection.

An additional property of this adder is that it detects overflow, also in an elegant manner with a single XOR gate, although it will take a bit more explaining as to why this is the case. In Chapter 1, we gave one rule for detecting overflow: if the most significant bit of the sum does not match both of the most significant bits of the addends, then an overflow is indicated. An alternative rule is the following: if the carry in for this bit does not match the carry out, then an overflow is also indicated. We can immediately see the equivalence of these rules with the following observation. The two cases where the sum does not match both bits in the addends are rows three and four in Table 4.1 (the count starts with row zero, as usual). In row three, we add 1 to 1 and get 0 (i.e., if this is the last bit we add two negative numbers and produce a positive). In row four, we add 0 to 0 and get 1 (i.e., if this is the last bit we add two positive numbers and produce a negative). However, these are also the only two rows where the carry in does not match the carry out. Case proved.

This new rule allows the construction of an overflow detector with a single extra XOR gate, shown on the right of Figure 4.6. The output of this gate will be 1 if and only if the carry in for the last bit, C_3, does not match the carry out for this bit, C_4. Thus, this gate will have a high output only when there is an overflow. This completes our initial survey of adder circuits. A full-fledged ALU of the sort we will consider in Chapter 7, will of course be more complex than the circuits we have discussed, but many of the most essential tricks of the trade are identical to those discussed here.

ON THE CD

System subtract on the CD-ROM illustrates the mechanism by which the circuit in Figure 4.5 works with 8-bit quantities. A negative number (B) is subtracted from a positive number (A). The positive form of B is placed at the top, in B(pos), and the system flips the bits and places this in B(neg). The rest of the system is identical to the ripple system except that the rightmost carry bit is constrained to be always on. This adds the extra 1 which is necessary to convert B into its negative form. The system shows the subtraction of 43 – 17, the result of which emerges after a few iterations of the animation.

DECODERS AND ENCODERS

In Chapter 1, we considered codes, or alternative ways of representing either the same quantity or the same item (such as a character). Often it is necessary to convert between codes, and the methods for doing so are considered here. By definition, if we are taking an *n*-bit code and converting it to an *m*-bit code, where $m > n$, then the process is known as decoding. Alternatively, if $n < m$, the process is known as encoding. Methods for decoding and encoding and some common applications for such are now considered.

Binary Decoders

The most common type of decoder is known as the binary decoder. This device has n input lines and 2^n output lines, and its task is to convert the binary number represented on the input lines to the quantity that this number represents. For example, Table 4.2 shows the truth table for the 2-to-4 binary decoder. Ignoring the EN signal for the moment and the top row of the table, and concentrating only on the inputs I1 and I0, we can see that the output line that is active corresponds directly to the binary quantity. For example, if the inputs are 1 and 0, representing the quantity 2, then the second output line, Y2, is triggered.

TABLE 4.2 The Truth Table for the 2-to-4 Binary Decoder

INPUTS			OUTPUTS			
EN	I1	I0	Y3	Y2	Y1	Y0
0	x	x	0	0	0	0
1	0	0	0	0	0	1
1	0	1	0	0	1	0
1	1	0	0	1	0	0
1	1	1	1	0	0	0

The EN signal is an enable line. Most composite combinational circuits provide one or more enable lines that allow the entire circuit to be disabled under the appropriate circumstances. Shortly, we shall see how this signal allows the construction of larger decoders from smaller ones. The truth table contains x's in the first row when the circuit is not enabled. These are don't care conditions; regardless of the values of the inputs, if the circuit is not enabled, then all of the outputs will be 0.

Figure 4.7(A) shows the circuit diagram for the 2-to-4 decoder corresponding to Table 4.2. Each of the outputs Y0-Y3 is preceded by an AND gate, which fires only when the enable EN is high and the I0-I1 signals are as indicated in the table. For example, the three inputs to the AND gate corresponding to output line Y2 are EN, I0', and I1. Thus, it will be active only when EN is high, when I0 is low, and when I1 is high.

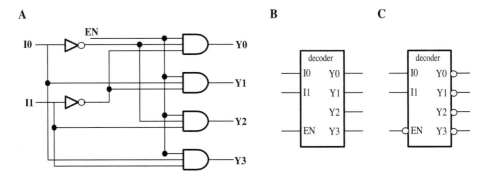

FIGURE 4.7 (A) The realization of a 2-to-4 decoder, (B) the logic symbol for this decoder, and (C) the logic symbol for a decoder with active low enable and active low outputs.

Typically, when illustrating a composite combinational circuit, such as a decoder, it is not necessary to provide all the circuitry as in 4.6(A); it suffices to provide a black box and the inputs and outputs as in 4.6(B). This simplification makes it easier to diagram circuits that use the decoder as a building block and also abstracts out the essential function of the device from its particular implementation.

TABLE 4.3 Truth Table for 2-to-4 Binary Decoder with an Active Low Enable and Active Low Outputs

INPUTS			OUTPUTS			
EN	I1	I0	Y3	Y2	Y1	Y0
1	X	X	1	1	1	1
0	0	0	1	1	1	0
0	0	1	1	1	0	1
0	1	0	1	0	1	1
0	1	1	0	1	1	1

We can also indicate with bubbles whether the input lines are active high or active low. Every circuit we have considered so far contains exclusively the former, but because it is cheaper to use NAND gates rather than AND, a more usual implementation of a decoder is symbolized in Figure 4.6(C). The inversion bubbles indicate that the enable is active low, and that all the outputs are active low. The circuit corresponding to this diagram is identical to that in 4.6(A) except that every AND gate is replaced with a NAND gate, and the enable line contains an extra inverter. The net effect of these changes is indicated in Table 4.3. In order to enable the decoder, it is necessary to set the EN line to low. Because all of the outputs are driven by NAND rather than AND gates, their values are simply the reverse of those in Table 4.2.

The notion of activity provides an additional layer of abstraction above the notion of 0s and 1s (corresponding to low and high voltages). It may be confusing at first but is worth mastering because it provides an extra degree of flexibility in one's design vocabulary. To say that an output is active means that it is doing something of significance from the point of view of the circuit's desired behavior. For example, in the case of the decoder, an active output means that this line is decoding the current set of inputs. Whether the line is then 0 or 1, or low or high voltage, depends on the implementation. If the line is active and we are using active high outputs, then it will be 1; otherwise it will be 0. Likewise, to say that an input line is active means that the circuit is responsive to this input. For example, in the case of the enable, it means that the circuit is operational. Whether we need to set this line to 0 or 1 for this to occur depends on whether the enable is active low, or active high, respectively. In summary, the notion of activity provides a way of talking about circuit inputs and outputs independently of the particular implementation. This effectively provides a clean functional description of circuit behavior without having to specify in advance the precise gates used in the final implementation.

System decode1 on the CD-ROM implements the 2-to-4 decoder in LATTICE. As in Table 4.2, when EN is off, the outputs are all 0. When EN is on, the active output corresponds to the binary quantity represented by the inputs. By clicking successively on the outputs Y0–Y3, you can observe that their defining truth functions correspond to the circuit in Figure 4.7. The variable definitions correspond to the three cells to the left of a given output unit (these shift from one output machine to the next because of the different positions of the input cells relative to the output). System decode2 contains an implementation of the active low output decoder with an active low enable corresponding to Table 4.3. The only difference between this and decode1 is that the enable has been negated, and the entire product function for each output has also been negated turning them into NANDs from ANDs. You should work through the four possible inputs and the two enable conditions for each system to confirm that the system works as expected.

While it is possible to build an arbitrarily large decoder with the method in Figure 4.7(A), at some point the fan-in will become prohibitively large. Thus, alternative methods must be sought; one such method is known as cascading. This is illustrated in Figure 4.8, which shows a 3-to-8 encoder constructed from two 2-to-4 decoders. It works as follows. When the most significant bit N2 is 0, then the top decoder is enabled (recall that this decoder is enabled by a low signal) and the bottom is disabled. Thus, in the first four rows of the truth table for this cascaded decoder in Table 4.4, the outputs Y4–Y7 are all 1 (i.e. inactive for this active low output decoder). Y0 through Y3 end up decoding input signals N0 and N1. Conversely, when N2 is high, then the top decoder is disabled and the bottom enabled. This results in the bottom four rows Y0–Y3 being inactive, and the top four rows Y4–Y7 decoding N0 and N1. The table shows the net effect of these operations: a diagonal row of 0s stretching from the top right to the bottom left, providing a binary decoding of the three input bits. One problem with the method illustrated in Figure 4.7 is that we have used the enable line for the third bit. Thus, if we wish to have an enable line for the new larger decoder it is necessary to provide an extra enable line on each of the 2-to-4 decoders that can be separately set.

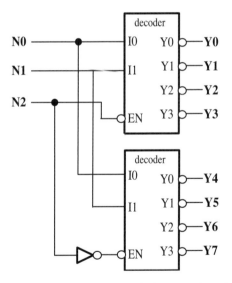

FIGURE 4.8 Two cascaded 2-to-4 decoders realizing a 3-to-8 decoder.

TABLE 4.4 The Truth Table for the 3-to-8 Cascaded Binary Decoder in Figure 4.7

INPUTS			OUTPUTS							
N2	N1	N0	Y7	Y6	Y5	Y4	Y3	Y2	Y1	Y0
0	0	0	1	1	1	1	1	1	1	0
0	0	1	1	1	1	1	1	1	0	1
0	1	0	1	1	1	1	1	0	1	1
0	1	1	1	1	1	1	0	1	1	1
1	0	0	1	1	1	0	1	1	1	1
1	0	1	1	1	0	1	1	1	1	1
1	1	0	1	0	1	1	1	1	1	1
1	1	1	0	1	1	1	1	1	1	1

Decoder Applications

Perhaps the simplest use of a decoder is to implement a two-level SOP expression in an alternative manner to that given in Chapter 3. This is illustrated in Figure 4.9, which shows the implementation of the function $PQ + QR \equiv \Sigma_{PQR}(3, 6, 7)$ with a 3-to-8 active-low output binary decoder. The key to understanding the operation of this circuit is the realization that each of the outputs Y0–Y7 will be active if and only if the corresponding minterm is represented by the inputs I0–I2. Thus, by merely taking the sum of the outputs, one obtains the sum of the minterms, which we know represents the function itself. The only minor complication arises if we are using a decoder with active low outputs, as is the case here. These must be inverted before they enter the OR gate; thus the bubbles where these lines enter this gate. Of course, we know from an earlier derivation that this is just a NAND gate.

In conclusion, to realize a canonical SOP function with an active high output decoder, it suffices to sum the outputs corresponding to minterm indexes with an OR gate; when using active low outputs, this summation is performed by a NAND gate. Please note however, that in general this will not be the most efficient method of realizing a function. First, we have not minimized the function, and thus we need to represent every minterm. Second, the decoder contains an AND (or NAND) gate for every minterm, when typically we want just a subset of the minterms. Thus, the simplicity afforded by the decoder in this capacity may be accompanied by a significant degree of waste (see Exercise 4.5).

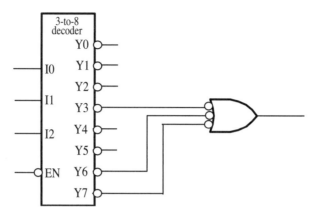

FIGURE 4.9 Realizing an SOP function with a decoder.

ON THE CD

System decode3 on the CD-ROM adds an additional function to the decoder in system decode1. This function, F, acts as an OR gate that is active whenever minterm 0 or minterm 2 are active (F is not set until the outputs Y0–Y3 are set; thus this simulation takes two time steps). This system is therefore a decoder realization of the SOP function $F \equiv I0'I1' + I0I1'$. System decode4 acts analogously with respect to the active low output decoder in system decode2. Note that the operator for F in this case is a NAND rather than an OR.

A somewhat more complex use of a decoder is illustrated in Figure 4.10. Suppose we are trying to drive the seven-segment LED display shown on the left, in the various configurations corresponding to each of the numbers 0 through 9, with the binary equivalents of these numbers. One way to do this is to harness the power of a 4-to-16 decoder to convert from the binary inputs to the numerical equivalent (the unused outputs 10–15 are not shown in the diagram). In this case, we will drive each of the LED segments separately, and let them be activated by the relevant inputs. For example, the top segment lights up whenever the binary input is equivalent to 0, 2, 3, 5, 7, 8, or 9. Thus, we sum these outputs and connect the output to this segment. The bottom left segment is driven by 0, 2, 6, or 8, and the corresponding outputs are accordingly summed. Similarly, we can drive all the other segments by their corresponding binary inputs, and the final result will be that the LED segments will display the numeral corresponding to these inputs (Exercise 4.6 asks you to drive two additional segments).

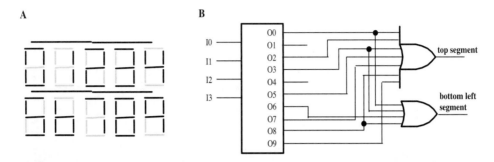

FIGURE 4.10 (A) A seven-segment LED display, and (B) the use of a decoder to drive the display for two of the segments.

ON THE CD

System decode5 on the CD-ROM is the original 2-to-4 decoder from system decode1 with the addition of three machines that respond when the top, middle, or bottom segments are active for binary inputs from 00 to 11. Each of the respective segment indicators is the sum of the possible inputs that trigger it. For example, when the inputs are both inactive, corresponding to the binary input 00 and the numeral 0, the top and bottom segments become active (see Figure 4.10(A)). Exercise 4.17 asks you to add machines that correspond to the remaining four segments.

Encoders

A decoder, by definition, transforms a smaller code to a larger one, and an encoder does the opposite. The simplest example of an encoder is the binary encoder, which transforms 2^n inputs and converts this to the corresponding binary code (this is the inverse operation of the binary decoder). This circuit could be used, for example, to transform an active input line among many such lines requesting interaction with a CPU into the binary code the CPU is expecting. Table 4.5 gives the truth table for the 8-to-3 binary encoder. Each of the eight possible input combinations is characterized by a single active value, and the outputs correspond to the binary equivalents of the line that is active.

The realization of this table follows directly by observing that each output corresponds to a sum of the four inputs when the output is active. Thus $N0 \equiv I1 + I3 + I5 + I7$, $N1 \equiv I2 + I3 + I6 + I7$, and $N2 \equiv I4 + I5 + I6 + I7$. These sums can easily be implemented with three OR gates, one corresponding to each of the sums.

TABLE 4.5 The Truth Table for an 8-to-3 Binary Encoder

INPUTS								OUTPUTS		
I7	I6	I5	I4	I3	I2	I1	I0	N2	N1	N0
0	0	0	0	0	0	0	1	0	0	0
0	0	0	0	0	0	1	0	0	0	1
0	0	0	0	0	1	0	0	0	1	0
0	0	0	0	1	0	0	0	0	1	1
0	0	0	1	0	0	0	0	1	0	0
0	0	1	0	0	0	0	0	1	0	1
0	1	0	0	0	0	0	0	1	1	0
1	0	0	0	0	0	0	0	1	1	1

NOTE

A good rule of thumb when implementing a truth function with multiple outputs is to implement each output individually, one at a time. This, for example, is what we have just done for the binary encoder. The only possible drawback of this procedure is that it ignores possible efficiency gains of the sort that we saw in the adder in which the sum and carry shared common circuitry. Still, as in this case, it is best to describe each output separately to begin with, and then to manipulate the resulting output functions algebraically to determine common subexpressions.

ON THE CD

System encode1 on the CD-ROM contains a 4-to-2 binary encoder. Inputs are arranged in a vertical bar, with I3 at the bottom and I0 at the top. The outputs N1 and N0 encode the inputs, assuming that only one input is active at a time (see the discussion below regarding the consequences of violating this constraint). Examination of the variable definitions for machines N1 and N0 reveals that they are the simple sums $N1 \equiv I3 + I2$ and $N0 \equiv I3 + I1$. These sums are derived in the same manner as in the 8-to-3 encoder binary encoder.

There are two ambiguities implicit in Table 4.5 that ordinarily means that we cannot get by with such a simple circuit. The first arises when every input is 0. In this case, the outputs would be all 0, but this is also the indication that I0 is active. The second ambiguity arises whenever more than one input is 1. For example, suppose both I1 and I2 were 1. Then the output would be $N0 = 1$, $N1 = 1$, and $N2 = 0$. However, this already corresponds to the case where $I3 = 1$. Some means of differentiating between these cases must be introduced.

TABLE 4.6 The Truth Table for a 4-to-2 Priority Encoder

INPUTS				OUTPUTS		
I3	I2	I1	I0	N1	N0	GS
0	0	0	0	0	0	0
0	0	0	1	0	0	1
0	0	1	X	0	1	1
0	1	X	X	1	0	1
1	X	X	X	1	1	1

Both of these problems are solved by what is known as the priority encoder. The truth table for the 4-to-2 version of this device (encoding the binary equivalent of 4 input bits) is shown in Table 4.6. The first thing you will notice is the addition of an extra output line GS, standing for "got something," to indicate whether any of the input bits are active. This allows the encoder to distinguish between the cases where all the inputs are inactive and the case where I0 is 1 and the rest of the inputs are inactive. The other, and most significant alteration is the creation of a priority scheme whereby only the most significant (leftmost in the table) is encoded. For example, row two of the table contains the sequence 001X for the inputs I3–I0. As before, the X is a don't care condition; regardless of whether I0 is 0 or 1, the output will be as indicated for this row. Likewise, rows three and four contain a similar priority scheme whereby the input lines to the right of the active bit are ignored. This table effectively disambiguates every case where more than one input line is active by paying attention only to the most significant active bit.

The procedure for generating the functions corresponding to the outputs relies on the fact that an X for a given input means that the variable ends up being factored out in the final expression. For example, row two can be represented as I3'I2'I1I0 + I3'I2'I1I0', but by factoring (or by observation on the Karnaugh map) this is I3'I2'I1(I0 + I0') ≡ I3'I2'I1. The same is true for multiple Xs in a given row; multiple factoring removes these variables in the product term. It then remains simply to sum the rows where the output is active to obtain the function for that output line. Thus N0 ≡ I3'I2'I1 + I3 (the sum of rows two and four), and N1 ≡ I3'I2 + I3 (the sum of the last two rows). By minimization (see Exercise 4.8), these can be reduced to N0 ≡ I3 + I2'I1 and N1 ≡ I3 + I2, respectively. Finally, GS is by definition active whenever at least one input line is active, and thus can be represented as the sum of this input, or GS ≡ I3 + I2 + I1 + I0.

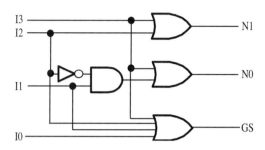

FIGURE 4.11 A realization of the priority encoder corresponding to truth table 4.6.

The realization of the functions for N1, N0, and GS can be achieved by any of the methods discussed in Chapter 3 for implementing a sum of product (SOP) expression. Figure 4.11 shows one such implementation with AND and OR gates. N1 is a simple sum and N0 is the sum of a product and the unaltered signal from I3. GS is the sum of all the inputs.

ON THE CD

System encode2 on the CD-ROM illustrates the priority 4-to-2 encoder. Inputs are as in encode1. However, in this system, more than one input can be active. For example, when I3 and I1 are active, then the system correctly encodes this by making both N1 and N0 active. Examination of these machines shows that they correspond to the output equations derived above for this encoder. This system also contains a machine for GS, which is active when at least one input is active. This disambiguates between the two input strings 0000 and 0001.

MULTIPLEXERS AND DEMULTIPLEXERS

Multiplexers and demultiplexers perform inverse operations. The former takes as input multiple streams of bits and selects one of those streams for the output. The latter takes a stream of bits as input and directs this stream to one of multiple outputs. Both act as digital switches, whereby the stream to switch on is controlled by one or more selection bits.

The operation of both is represented in a hypothetical MUX-DEMUX circuit in Figure 4.12. The MUX selects among four sources, according to the select bits, and puts the selected stream on the bus (a bus is simply a collection of related data lines). In this case, the selection bits indicate that source 01, or 1 should be chosen. Once on the bus, the data stream is then routed to one of four outputs, in this case destination 11 or 3. In this manner, a data stream can be routed from 1 of n sources and directed to 1 of *n* outputs. Note that we have not specified the size of the stream. This could be a single bit, a byte, or multiple bytes depending on the size of the application. Also, note that by successively stepping through the selection bits to the MUX, we can place different bit streams on the same bus. This common multiplexing operation allows multiple input lines to be reduced to a single line by having the signals occupy different time slices, and is useful whenever a trade-off in speed versus line count is warranted.

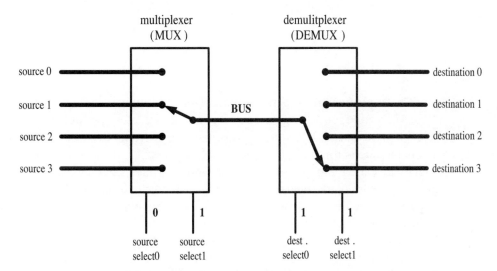

FIGURE 4.12 A mutiplexer-demultiplexer combination that selects from one of four sources and directs the selected source to one of four outputs. Selection bits are 01 for the source and 11 for the destination.

Multiplexers are characterized by both the size of the stream for each input source and the number of such streams. In general, 2^n streams require n selection bits, with each combination of these bits addressing one of the possibilities. Let us consider the implementation of the MUX in Figure 4.12, in which, for simplicity, the size of each of the sources is 1 bit wide. This is designated as a 4-input, 1-bit mulitplexer, and its truth table is given in Table 4.7. When the enable is inactive, the output is also inactive. When this is not the case, the output will reflect whatever is on the selected input line. This table is different from those we have seen previously in that the output is not a fixed value, but reflects the activity on the selected line.

102 Digital Design: From Gates to Intelligent Machines

TABLE 4.7 The Truth Table for a 4-input 1-bit Multiplexer with Active High Enable

EN	S1	S0	OUT
0	x	x	0
1	0	0	I0
1	0	1	I1
1	1	0	I2
1	1	1	I3

The implementation of the 4-input 1-bit multiplexer is illustrated in Figure 4.13. There are four AND gates, corresponding to the four possible inputs, and the select bits S0 and S1 select the appropriate gate based on their values. In addition, an enable line is routed to each of the AND gates. As discussed in Chapter 3, an AND gate can be thought of as a filter with two lines, data and control. It propagates the values on the data line through the gate when the control line is 1, and otherwise the gate produces an output of logic 0. This implementation draws on this analogy; the difference is that now we have multiple conditions for the control lines. The control for each of the successive AND gates corresponds to the conjunction of the select bits for that gate and the enable line. When these are all logic 1, then the gate allows the indicated input to go through unhindered.

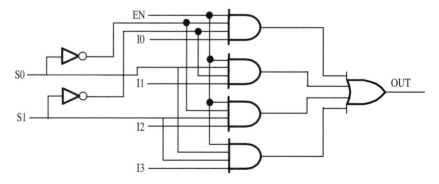

FIGURE 4.13 A realization of a 4-input 1-bit multiplexer with active enable.

As an example, consider the case where S0 is 0 and S1 is 1, and the enable is also 1. Only the second AND gate from the bottom will potentially be active, and only when I2 is active. In effect, this AND gate will transfer to the output the value of I2, and as Table 4.7 shows, this is precisely what we need it to do.

Table 4.8 shows the truth table for an 8-input 1-bit multiplexer. If there are eight inputs, then there must be $\log_2 8 = 3$ selection bits to determine which of the eight lines is selected. To implement this mulitplexer, we can use the same strategy as in Figure 4.13. Each of the AND gates is fed by the combination of selection lines that select for this gate. There will be a total of eight such gates, one for each output possibility, and as before, the enable signal feeds into each of these (Exercise 4.7(a) asks you to complete this realization).

TABLE 4.8 Truth Table for an 8-Input 1-Bit Multiplexer with Active High Enable

EN	S2	S1	S0	OUT
0	x	x	x	0
1	0	0	0	I0
1	0	0	1	I1
1	0	1	0	I2
1	0	1	1	I3
1	1	0	0	I4
1	1	0	1	I5
1	1	1	0	I6
1	1	1	1	I7

Multiplexers typically route streams that are larger than one bit wide. Table 4.9 shows the truth table for a 4-input 2-bit multiplexer. Note that there are two outputs, labeled A and B, one for each of the bits in the selected bit stream. The total number of inputs will be $4 \times 2 = 8$. That is, there will be a total of four streams, selected by $\log_2 4 = 2$ select bits, with each stream consisting of a parallel input of two bits. To implement a multiplexer such as this, we simply duplicate the 1-bit circuit the relevant number of times. In this case, this will entail two copies of the circuit in Figure 4.13, each of which will be fed by the relevant bit from each of the four input streams (Exercise 4.8(b) asks you to fill in the details).

TABLE 4.9 Truth Table for a 4-Input 2-Bit Multiplexer with Active High Enable

EN	S1	S0	OUT A	OUT B
0	x	x	0	0
1	0	0	I0A	I0B
1	0	1	I1A	I1B
1	1	0	I2A	I2B
1	1	1	I3A	I3B

ON THE CD

System MUX1 on the CD-ROM illustrates the operation of a 2-input 1-bit multiplexer. There are two total streams, each 1 bit wide. These are indicated in the system by the machines I0 and I1, which shift their bits one cell to the right on each simulation step. The output computes its value according to the function $F \equiv (ENS0'I0) + (ENS0I1)$. That is, if the circuit is enabled and the select signal S0 is 0, then stream I0 is selected. When the circuit is enabled and S0 is 1, then stream I1 is selected. The output machine then places its result on the bus, which also shifts the cells right with each animation step. In the current configuration with S0 = 1, the I1 stream is placed on the bus. Thus, every third cell is activated as the animation is stepped through. Turning off S0 selects stream I0 and two of every three cells then become active on the bus. This system illustrates that multiplexers are typically dynamic devices; that is, rather than merely computing truth functions, they act as routers for dynamic input streams.

System MUX2 illustrates another use of the multiplexer. In this system, two 1-bit input streams traveling at half speed are entering a multiplexer (the streams are designed to advance only when the blink machine is on). The multiplexer is as before, but the selection bit is coordinated with the blink. This means that every other time step the multiplexer selects from the top stream and on the other half of the cycle selects from the bottom stream. The net effect of this operation is to interleave the two streams onto the bus, which operates at normal speed (here, the streams 11011 and 01010 are combined to yield 0111001101). This operation, in which multiple input streams traveling at slower speeds are woven into a single output stream traveling at a higher speed is known as TDM (Time Division Multiplexing). TDM is useful whenever it would be too costly to have a number of slow direct lines between points, and it is better to have a single high-speed line, which is then "demultiplexed" at the destination (see Exercise 4.21).

An additional use of a multiplexer is to implement a general logic function. Figure 4.14 illustrates how this can be accomplished. Suppose the minterms are as indicated in the Karnaugh map in the left of the figure. Then a 4-input 1-bit multiplexer can realize this function as follows. Let the selection bits be two of the variables in the function, say A and B. Then each of the four permutations of these bits, 00 through 11, will select one column of map. For example, when $S_1 = A = 0$ and $S_2 = B = 0$, the first column of the map will be selected. It then suffices to realize the indicated values of the variables C and D on the selected line, in this case, the topmost line. From the map we can see that the first column can be represented by the variable D, so this is the input on this line, as can be seen in the circuit diagram on the right. Likewise, when $A = 0$ and $B = 1$, the column is empty, so we place 0 on the second line from the top. The third line corresponds to the case where $A = 1$ and $B = 0$, which as shown on the map corresponds to a single minterm in which $C = 1$ and $D = 1$. Thus, we need an extra AND gate to represent this column. Finally, when $A = B = 1$, the function is 1 everywhere, so we place a 1 on the fourth input to the MUX. Similarly, any four-variable function can be implemented. Please confirm that that output of the multiplexer is correct for each of the 16 input possibilities.

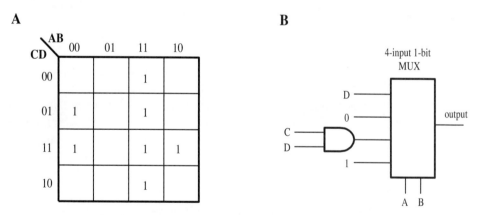

FIGURE 4.14 Realizing a function of four variables with a multiplexer. (A) The Karnaugh map for the function, and (B) the realization of the function by a 4-input 1-bit multiplexer.

We now come to the shortest description of a complex circuit in the book, that of the demultiplexer. It receives this honor because it is, in fact, identical to a circuit that we have already described, the decoder, with one small change. In Table 4.8, we have simply redescribed Table 4.2, the truth table for the 2-to-4 decoder with enable, by taking the enable bit and placing it in the output, and relabeling the input bits as select bits. Recall that when the decoder is enabled, it will activate the line indicated by the inputs, and when it is not enabled, all outputs will be inactive. Thus, the selected output will always reflect the state of the enable. For example, if the inputs S1 and S0 are 1 and 0 respectively, then Y3 will be identical to the enable line. If we then merely relabel the enable as the input line, we have created a 4-output 1-bit demultiplexer, that is, a circuit that routes the bit on this line to one of the outputs based on the selection bits.

TABLE 4.8 Truth Table for a 2-to-4 Decoder Recast as a Truth Table for a 4-Output 1-bit Demultiplexer

INPUTS		OUTPUTS			
S1	S0	Y3	Y2	Y1	Y0
0	0	0	0	0	EN
0	1	0	0	EN	0
1	0	0	EN	0	0
1	1	EN	0	0	0

Identical truth tables have identical implementations, and therefore a decoder and a demultiplexer are the same (and, in fact, chip manufacturers describe this circuit as a decoder/demultiplexer). To implement a demultiplexer that chooses among a larger number of outputs, it suffices to use a binary decoder with more inputs, and to implement a demultiplexer that routes a larger number of bits, multiple decoders of the appropriate size may be used. As in the case of cascading decoders, it is necessary to provide at least one additional enable line because the first enable line is being used for data.

ON THE CD

System DeMUX on the CD-ROM is identical to decode1 with two exceptions. First, the inputs are relabeled as select bits. Next, there is a shift-right line that feeds the enable cell and a shift-right line extending from each of the outputs. As the simulation is stepped through, each successive bit in the input line is placed in the enable cell. Then, depending on the select bits, this input stream will be routed to the appropriate output. In the default configuration for this machine, S1 is active and S0 inactive. Thus, line Y2 is selected for, and the inputs are routed to this cell. Examination of the truth functions for each of the outputs Y0–Y3 show that they are identical to that of the decoder; that is, each is a product of the enable and the corresponding combination of select bits that decode this line.

PROGRAMMABLE LOGIC DEVICES (PLDs)

Early on in the development of digital design, it was realized that it would be useful to have circuits that could be programmed to achieve various logical functions rather than having to construct the device from scratch. The result, PLD (programmable logic device) may be thought of as the hardware analog of a programming language. In the latter paradigm, you are not supplied with a running program, but with the tools with which to create programs; likewise PLDs provide the means to realize combinational designs without indicating in advance the nature of those designs. Large-scale prototyping of integrated circuits is accomplished with FPGAs (field-programmable gate arrays), which are larger and more complex versions of PLDs.

PLDs and FPGAs represent an intermediate level between design using elementary gates and the construction of an application-specific integrated circuit (ASIC) to achieve some task, and are useful when navigating between the awkwardness of the former and the expense of the latter. Once a design is decided upon, and especially if the device is expected to be manufactured in large numbers, a more efficient, but initially more costly, ASIC may be substituted for the PLD or FPGA in question. Three types of PLDs will be considered here: programmable read-only memory (PROM), programmable array logic (PAL), and the programmable logic array (PLA).

Programmable Read Only Memory (PROM)

As its name suggests, a PROM is a type of memory that can be externally set. A circuit for achieving this result is shown in Figure 4.15, which illustrates a ROM that can hold 16 memories each a byte wide. The schematic shows that it accomplishes this with a 4-to-16 decoder whose outputs are potentially connected to eight OR gates. The PROM originally comes with no connections made, but after programming, the outputs of the decoder are connected to the OR gates as indicated by the Xs. We will not worry about the physical details about how such a connection is made. Suffice it to say that in a PROM, these connections are not hard-coded but can be externally set (and if the device is an EPROM, or erasable PROM, they can be set and reset up to 10,000 times). As with all the PLDs discussed here, the X symbolizes that the given line has been fused, and all other lines are assumed to be not fused, or not connected. For example, in the case illustrated in Figure 4.15, the OR gate serving output Y7 effectively has become a 4-input OR gate, with the inputs the 0, 1, 12, and 14 output lines of the decoder.

The PROM works as follows. Suppose one wishes to get the contents of a given memory, say the 14th item. One addresses this memory location by providing the decoder with the binary representation of this address, 1110. This will then make the 14th output line of the decoder (and no other lines) active. If the PROM

108 Digital Design: From Gates to Intelligent Machines

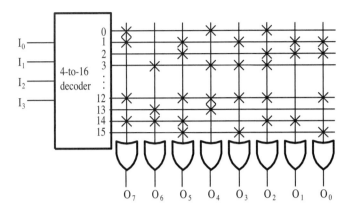

FIGURE 4.15 Realizing a PROM with a decoder, OR gates, and fusible connections.

has been set as in Figure 4.15, then outputs Y7, Y6, Y5, Y2, and Y1 will be active. In effect, the input address 1110 will recover the byte-length memory 11100110. Similarly, each of the 15 other memories can be retrieved. Larger memories can be achieved with bigger decoders, and more bits per memory can be achieved by adding OR gates. Memory, of course, is an important component of any computer, and it will be treated in more detail in Chapter 7.

ON THE CD

System PROM on the CD-ROM contains a small memory system that stores two bytes. The select bit to the left selects either the bottom or top memory depending on whether it is active or inactive, respectively. If the former, as in the default configuration, the row of cells labeled "(1)" are activated. The activation of this row places the byte stored above on the out set of cells. Successively clicking on the forward button triggers this process. Setting the select cell to off reads the other memory, in "(0)" onto the output. Note that this is an illustration of the workings of a PROM, rather than a strict simulation. In an actual PROM, the current flows from the decoder onto the selected line, and then through the fused connections, a process that is not easily simulated in LATTICE. Later, we will explore easier ways to read from and write to memory with state-based systems.

Programmable Array Logic (PAL®)

A PAL, unlike a PROM, realizes a set of SOP representations rather than simple sums. It accomplishes this with a fixed number of AND gates feeding into a set of OR gates, with each OR gate corresponding to a single function. The device is programmed by fusing connections, as in the PROM. Figure 4.16 shows how this is carried out in the case in which we wish to realize the functions $F_1 = ABC + A'D'$

+ B'CD' and F_2 = ACD' + B'D. In the illustrated case, each OR gate is fed by three AND gates. Both F_1 and F_2 consist of three or fewer product terms, so this PAL will be adequate to represent these functions. The functions are realized by fusing the connections indicated to represent the products. For example, the first term in F_1 is formed by programming the input lines for A, B, and C to connect to the first AND, and other terms are likewise formed. Note that it is not necessary to connect every variable or its complement to each AND gate, nor must every AND gate be used if the number of terms is less than the number of such gates, as in F_2. If more product terms are necessary, a PAL with greater numbers of gates per function must be used. PALs typically come with the capacity to represent greater than two functions, and may be augmented by enables on the outputs.

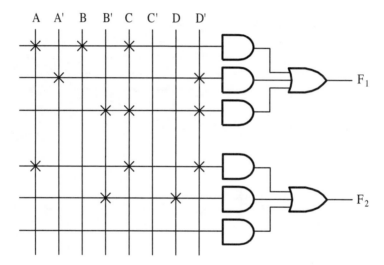

FIGURE 4.16 A PAL realizing two SOP functions.

Programmable Logic Array (PLA)

A PLA is a more general form of a PAL in which the inputs to the OR gates can be programmed as well as the inputs to the AND gates. While it is always possible to realize an arbitrary SOP function with a PAL with sufficient AND gates, the PLA allows economies to arise by reusing the outputs of the AND gates. Figure 4.17 shows a characteristic example where this occurs. The goal is to represent the following functions:

$$F_1 = A'B' + ABC + A'CD$$
$$F_2 = A'B' + C'D' + AC' + CD'$$
$$F_3 = A'CD + AC'$$

This is done in two steps. First, the product terms are formed by fusing the appropriate connections. For example, the leftmost product term, P1, is A'B'. Next, the products are connected to their respective functions. For example, to form F1, the first three products terms are used.

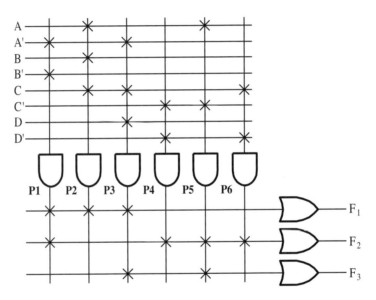

FIGURE 4.17 A PLA with four input variables, six product terms, and three output functions.

While there is nothing complicated about programming a PLA, there *is* one aspect of this process that makes it different from designing for separate SOP functions, one at time. Note that the pictured PLA is only capable of representing six products, but that the total number of product terms for all three functions is nine. The PLA is able to represent these because product terms P1, P3, and P5 are used by more than one function. In order to achieve this, the product terms were chosen not just to minimize the function in question, but also with a view toward their reuse by other functions. For example, F_1 could also be represented by the equally complex A'B' + ABC + BCD (see Figure 4.18). But then the final product term appearing in the original formulation, A'CD, would not be able to be reused by function F_3.

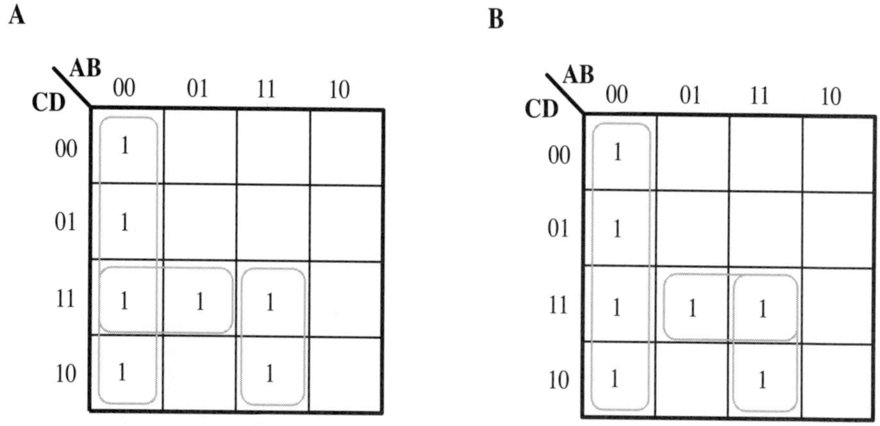

FIGURE 4.18 (A) The original representation of the function F_1, and (B) an alternative representation with an equal number of products.

Designing to minimize the total number of terms in multiple functions is a special case of design with a complex set of objectives. In general, such problems are intractable, that is, it is not possible to generate an optimal solution within a reasonable period. To see this, let us formalize the problem by introducing the notion of an objective function, or a function that we are attempting to minimize. A typical example of the kind of function we would be interested in is

$$FObj = 1.0 \times \#gates + .2 \times \#inputs, \tag{4.5}$$

where #gates is the total number of gates in the design and #inputs is the total number of input lines over all gates.

It would be nice if we could follow a simple rule or set of rules that continually reduces F_{Obj} until a minimum is achieved. Figure 4.19 shows why this is not possible in general. In this diagram, the design is at the location of the ball, and F_{Obj} is shown as a function of the current design choices (in this simplified case, we can go in one of two directions, left or right). If we simply try to minimize this function, we will end up rolling down the hill and we will achieve a better design. However, as the figure shows, we have entered a local minimum rather than a global one. There is a better design over the hill, but in order to get there, we would first have to make the design worse. But this entails a more global search process rather than a simple local one, and this is by necessity more difficult and takes more time.

We saw an example of this phenomenon in Chapter 2. As you may recall, in order to simplify a function (make it smaller), we first had to take the counterintuitive step of expanding it. In other words, first we had to make it worse according to the

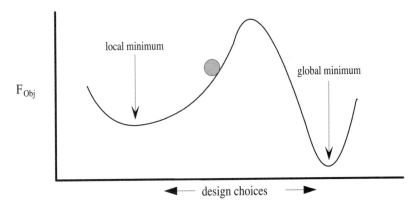

FIGURE 4.19 Searching for a minimal objective function when multiple minima are present.

objective function before we made it better, and therefore an algorithm that continuously tried to reduce the value of this function would not work.

If it is not possible to generate an optimal solution in a reasonable amount of time, what is the alternative? We aim for an approximate solution that is good enough. The key to generating such solutions lies in both the appropriate use of optimization software, and often the intuition of the designer, who is able in advance to prune the types of solutions from the space of possible solutions that are unlikely to work well. Once again, it is apparent that good design is as much an art as a science.

SUMMARY

It would be foolish in the extreme for a house builder to make his own wood beams from raw lumber, to bake his own bricks, and to stretch copper into wires for the electric lines. Likewise, when constructing a complex digital circuit, it is often best to begin with larger building blocks than simple gates. The first example of a building block that we examined was the adder. We saw that addition was a disguised form of logic, and therefore could easily be implemented with a digital circuit. We also saw how to propagate the result of one column to the carry bit of the next column with the ripple-carry adder. We then examined how a space-time trade off may be affected by the use of a carry–look-ahead scheme. Finally, we considered how two's complement addition could be efficiently implemented with simple gates. Adders such as these are important components of the Arithmetic Logic Unit of a computer, a topic that we will cover in more detail in Chapter 7.

Next, we considered decoders and encoders. The former, by definition take an n-bit code and transform it into an m-bit code, where m > n, and the latter, do the opposite. The most common type of decoder is the binary decoder, which takes n inputs and activates the line (out of 2n output lines), which corresponds to the binary quantity represented by the input. Binary decoders can be cascaded to make large decoders by a clever reuse of the enable line as an extra input bit. The most common type of encoder also involves binary conversion. However, in this case 2n inputs are converted to an n bit output. In addition, it is typical to have a priority scheme, whereby the most significant active bit in the input is the one that is encoded by the output. The realization of both decoders and encoders follows from an examination of each of the output columns, one at a time. A separate circuit is constructed for each output by invoking the usual bag of tricks involved in taking a set of minterms and converting it into a digital realization.

Multiplexers and demultiplexers form another type of complex digital circuit. Multiplexers take a number of streams of parallel data and switch one of these streams onto a bus. The selection bits on the multiplexer determine which stream is chosen. Multiplexers may also be used to implement a general logical function. Demultiplexers perform the inverse operation. They take a single stream and route it to one of many outputs. The realization of a multiplexer is straightforward and is a matter of allowing the select bits to act as the control line on an AND gate. The realization of a demultiplexer is even simpler. A demultiplexer and a decoder have identical truth functions if the enable line of the decoder is used as the input. To demultiplex a stream with multiple bits, multiple decoders can be used.

Every circuit we have seen to this point performs a fixed function or set of functions. In contrast, PROMs, PLAs, and PALs are more general-purpose devices that allow some degree of functional programming. PROMs allow memories to be stored with the use of fused connections between a decoder and a set of OR gates. PLAs allow the fusing of connections on AND gates to achieve SOP representations of multiple functions. PALs provide the greatest amount of flexibility by allowing both the fusing of connections on AND gates and OR gates. In many cases, this results in design efficiencies because product terms can be reused in other sums. All of these programmable devices represent a foreshadowing of the most general programmed device, the computer, which we will consider in extensive detail in Chapter 7.

EXERCISES

4.1 Confirm with a truth table that function (4.1) is equivalent to (4.1') and that (4.2) is equivalent to (4.2'). (These equivalences allow the construction of the full adder shown in Figure 4.2.)

4.2
  (a) Show the truth table for the addition of two 2-bit numbers (your table will have four columns for the inputs, two for each number, and three outputs, two for the sum and 1 carry out).
  (b) Suppose you were to create the truth table for 4-bit addition in the same way. How many rows would the table have? How about for 8-bit addition?
  (c) Given your answer above, why is binary addition accomplished with a ripple-carry adder or the equivalent sequential process rather than the direct implementation of the truth table for the addition?

4.3 Compute C_{i+3} in terms of C_i using recursion relation (4.3').

4.4 Justify Figure 4.5. That is, why in these instances will the carry for the column be 1 with the two previous columns as specified?

4.5 Realize the following functions with active, low output decoders, and in both cases indicate how many gates are wasted relative to an implementation using NAND gates alone.
  (a) $\Sigma_{PQR}(1,3,5,6)$
  (b) $\Sigma_{ABCD}(4,6,7,13,14)$

4.6 Use a decoder to drive the bottom and bottom right segments in the seven-segment LED display.

4.7 Use four 3-to-8 decoders and one 1-to-2 decoder to create a 5-to-32 cascaded decoder.

4.8
  (a) Implement the 8-input 1-bit multiplexer corresponding to the truth table in Table 4.8.
  (b) Implement the 4-input 2-bit multiplexer corresponding to the truth table in Table 4.9.

4.9 When producing the functions for the realization of the 4-to-2 priority encoder, we claimed that I3'I2'I1 + I3 was equivalent to I2'I1 + I3 and that I3'I2 + I3 was equivalent to I3 + I2. Show that these equivalences hold with four-variable Karnaugh maps.

4.10
  (a) Produce the truth table for an 8-to-3 priority encoder.
  (b) What are the functions (unsimplified) corresponding to the outputs of the table?

4.11 Give the expression for function F in Figure 4.20 assuming that each block represents a 2-input 1-bit multiplexer and the select line enters at the bottom of the block. (Hint: First figure out the general function for such a MUX.)

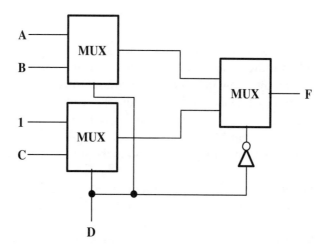

FIGURE 4.20 The circuit diagram for Exercise 4.11.

4.12 Show the truth tables for the following devices:
   (a) An 8-input 1-bit multiplexer
   (b) A 4-input 2-bit multiplexer
4.13 Reimplement the encoder in Figure 4.11 with NAND gates.
4.14 How many total input lines does an n-input m-bit multiplexer have? How many selection lines would be necessary?
4.15 Enumerate the memories encoded by the PROM in Figure 4.15.
4.16 List the functions encoded by the PAL in Figure 4.21.

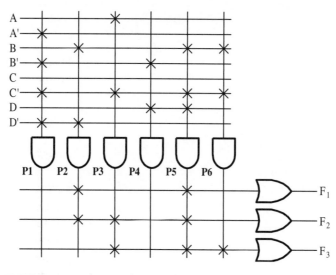

FIGURE 4.21 The PAL for Exercise 4.14.

LATTICE Exercises

4.17 Create a 4-bit adder along the lines of look-ahead in which the last two carry columns are computed with the look-ahead function.

4.18 Add an overflow indicator to system subtract.

4.19 Modify system decode5 with additional cells corresponding to the four LED segments not already present in the system.

4.20 Modify system MUX to create a 2-input 2-bit multiplexer. (Hint: Create two copies of the existing system, but make sure that the select bit is the same for both.)

4.21 Modify system MUX2 by adding a demultiplexer that takes the full-speed interleaved signals on the bus and returns them to two half-speed separate signals.

4.22 Modify system PROM so that each memory is 2-bytes wide rather than a single byte (you will need to change the cell size from coarse to medium).

5 Elements of Sequential Design

In This Chapter

- Introduction
- Latches
- Flip-Flops
- Registers
- Summary
- Exercises

INTRODUCTION

Combinational circuits, no matter how complex, produce outputs that are strict functions of the inputs. But it is easy to see that this is not all we would want from a circuit. Most devices of any complexity respond in a way that is a function of not only the inputs to the device, but also to the state of the device. For example, because of the limited number of keys, a cell phone often makes multiple use of its inputs. A given key may act in one mode to enter a character, and in another to lock or unlock the keyboard. People are a much more interesting example. We take it for granted, but people respond in quirky and idiosyncratic ways, even when we hold the stimulus constant. One day we may meet Harry on the street and after we say hello, he may boisterously respond with a discussion of his recent vacation; the next day, under identical circumstances, he may complain about his boss. In other words, to predict human behavior it is not sufficient to know the current context and the personality of the person in question; one must also know the internal state of that person. Of course, we only have indirect access to this state, through facial expressions and the like, or previous behavior, making the human system at best semipredictable.

Although any given individual is unpredictable, this does not necessarily mean that the sum total of human behavior contains the same degree of unpredictability. The idea is that the quirks of individual behaviors average out, resulting in the approximately predictable net behavior. For example, you may be excited about a stock that you are about to purchase, but then you postpone your purchase because your leaky roof puts you in a foul mood. Across the country, however, someone who was less enthusiastic about the stock is feeling particularly ebullient (his roofer just gave him a bill that was less than he expected). The net effect of these local fluctuations is to cancel out, and the stock achieves the price it would have anyway.

An extreme form of this view was expressed by the great science fiction writer Isaac Asimov in his Foundation trilogy. Asimov suggested that history could be made into a science, a field he dubbed psychohistory. According to this view, the unfolding of history is inevitable. It may seem to be the collection of actions of powerful leaders, but these leaders are in fact being driven by forces beneath the surface that may be unknown even to them. The most important of these forces is the collective behavior of the people that elected (or indirectly appointed these leaders). This behavior, the hypothesis states, is unpredictable at the atomic level (i.e., the level of the individual) but is law-like in the aggregate.

None of these hypotheses, especially the strong form appearing in Asimov's writings, have been confirmed. It must be said, however, neither have they been extensively investigated. Moreover, the idea of collective intelligence is just coming to the forefront, and it could be that a fuller understanding of this topic could help advance some form of this theory. This discussion will resume in Chapter 9, when we consider the notion of emergent behavior.

We will not begin to treat the implementation of intelligent behavior yet. Suffice it to mention that it is not a mere matter of hooking up a few gates in the right way. Nevertheless, the fundamental point remains: to achieve complex behavior, we need some way of holding a state, and have the device respond as a function of that state as well as the inputs to the device. In other words, we need a system with memory. This chapter will discuss the elements that will allow us to achieve this goal, and Chapter 6 will show how these elements can be put together to realize state-dependent devices.

LATCHES

Combinational circuits have no memory because electrical current courses through them from inputs to outputs without interruption. Consider, however, if at some point in the circuit, a signal is fed back to an earlier part of the design. If this is done in the right way, then current will flow in a circle, maintaining an earlier state of the device. This is the inspiration for the latch. Here we will consider two types: the SR latch and the clock-triggered D latch.

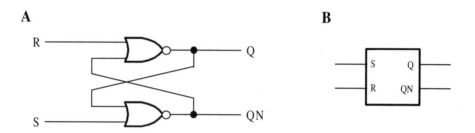

FIGURE 5.1 (A) The SR latch design and (B) the logic symbol for this latch.

SR Latch

The circuit for the SR (set-reset) latch is shown in Figure 5.1(A) and the logic symbol is shown in Figure 5.1(B); the truth table is given in Table 5.1. We first consider how each row of the truth table follows from the circuit design. In the following analysis, we assume that the natural states for Q and QN are to be complementary states, that is, $Q \equiv QN'$ (although this is not a strict requirement, as indicated in the last row of the table). The top row of Table 5.1 represents a condition which we have not yet seen, and is called the maintain or "memory" state. When both the set and reset lines are low, the last values of Q and QN remain at their previous values.

To see this, let us examine the equations for the outputs:

$$Q^+ \equiv (R + QN)' \tag{5.1}$$

and

$$QN^+ \equiv (S + Q)', \tag{5.2}$$

where Q^+ is defined as the next state of Q and QN^+ is the next state of QN. These equations follow directly from the circuit design. Note that Equations 5.1 and 5.2 represent a departure from the normal logical equivalences that we have been using so far in that the left-hand side represents the next state of the indicated variable. The right-hand side is the current state of the indicated variables. The distinction between current and next states is necessary whenever we have a circuit with a feedback connection.

TABLE 5.1 The Truth Table of the SR Latch

S	R	Q	QN
0	0	last	last
0	1	0	1
1	0	1	0
1	1	0	0

Returning now to the top row of Table 5.1, let us examine how the maintain state works. If Q and QN are in complementary states (as they will be ordinarily), then when R = S = 0, $Q^+ \equiv QN'$, and $QN^+ \equiv Q'$ by Equations 5.1 and 5.2 respectively, and no change will occur.[1] The next state of Q is the complement of QN, that is, Q stays in the state it is already in, and the next state of QN is the complement of Q, and QN also remains in its current state. If Q = QN = 0 (an unusual but possible state), then one of the two will be set to 1, depending on which changes first, and the other will be set to 0. Likewise, when Q = QN = 1, one will go to 0 first, and the other to 1. In these cases, if both Q and QN change at once, the system will go into an oscillatory mode (see the timing discussion below). However these are unusual cases, which we attempt to avoid when working with latches. Thus, the top row of the table can rightly be called a maintain state.

It would be useless to simply maintain a state without being able to set it, and this is the purpose of the next two rows of the table. When S = 0 and R = 1, Q⁺ = 0, by Equation 5.1, and therefore QN⁺ = Q' = 1, by Equation 5.2. This is known as the reset state. Alternatively, when S = 1 and R = 0, then QN⁺ = 0 and Q⁺ = QN' = 1 by the same governing equations. Finally, we consider what happens should S and R both become 1. In this case, Q⁺ = QN⁺ = 0. Ordinarily, we would not use this set of inputs because we desire that Q and QN are complementary, but we may not always be setting R and S explicitly (they could be the outputs of another circuit, for example), so we report this case for reference.

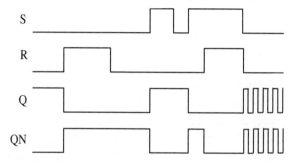

FIGURE 5.2 A typical timing diagram for the SR latch.

Figure 5.2 elucidates the operation of the SR latch by showing a typical timing sequence. Unlike the timing diagrams in Chapter 3, for the purpose of simplicity, we will assume that the inputs to the latch have an immediate effect on the outputs, and that the rise and fall trajectories for all signals are purely vertical. Each signal can be in one of two states: high (or 1), corresponding to the upper horizontal line, and low (or 0) corresponding to the lower horizontal line. The initial state of the circuit has Q at high and QN at low. When R goes to high, this situation is reversed, and then reversed twice as S goes high and R goes high again. When both S and R become high, both Q and QN go low, as indicated in Table 5.1.

Then something unusual happens when S and R, by coincidence, drop to 0 nearly simultaneously. By Equation 5.1, $Q^+ = QN'$, but QN = 0 and therefore Q goes to 1, and by Equation 5.2, $QN^+ = Q'$, but Q also is 0, and therefore QN also becomes 1. On the next time step, Q and QN become 0; this also follows directly from the governing equations and the values of R, S, Q, and QN. But now we are back to the state where all the values are 0, and so on the next time step Q and QN become 1, and so forth. The net effect is an oscillation between high and low states for both of the outputs, and no stable state is reached. It is relatively difficult to achieve this sort of instability, as under most circumstances, Q or QN will be set first to a particular value and force the other into the complementary state. We will not treat this type of oscillatory behavior further in this chapter, but simply note that it is always a possibility once feedback is allowed, and therefore the conscientious designer must be aware of this eventuality, however rare it may be.

ON THE CD

System SRlatch in the Chapter5 folder on the CD-ROM illustrates the behavior of this device. There are four cells, corresponding to R, S, Q, and QN. R and S can be set as desired, and Q and QN are governed by Equations 5.1 and 5.2 respectively. First setting S and R to 0, and Q to 1 and QN to 0 we can see that there is no change in these values after stepping through the simulation. Likewise, with these settings of S and R, if Q = 0 and QN = 1, then these values are also maintained. When R = 1 and Q = 0 (the reset case), Q goes to 0 and QN goes to 1 in at most two time steps, regardless of the initial values of these cells. When Q = 1 and R = 0 (the set case), Q goes to 1 and QN goes to 0. The last row of the truth table in Table 5.1 can be simulated by setting both R and S to 1. In this case, both Q and QN drop immediately to 0, regardless of their initial conditions. There is one final case to be considered, that corresponding to the oscillation on the right of Figure 5.2. If we set all four cells to 0, then this condition is generated, as can be seen by either successively pressing the step button, or hitting the run button once. This occurs because from the point of view of the system, both S and R have dropped to 0 simultaneously, as these settings have been made when time is not running in the simulation. Note that this condition is harder to generate with a real SR latch because typically either S or R will drop first, triggering one of the conditions corresponding to the two middle rows of Table 5.1.

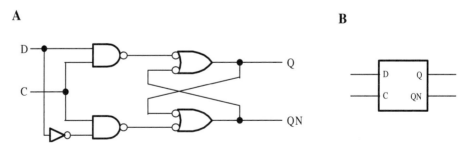

FIGURE 5.3 (A) The D latch design, and (B) the logic symbol for this latch.

D Latch

The D latch, shown in Figure 5.3, overcomes a key shortcoming in the SR latch. It is awkward to have to use two lines to set the latch when only one bit is being set. It would be more natural to have a single line to set the latch, and another to tell whether setting or maintaining is required. This is the method of the D latch, the truth table for which is given in Table 5.2. The D line contains the bit to set the latch, and the C line enables the set (it is called C rather than EN because we will use it for a clock signal when forming the D flip-flop, in the next section).

TABLE 5.2 The Truth Table for the D Latch

C	D	Q	QN
1	0	0	1
1	1	1	0
0	x	last	last

As before, we prove that the circuit realizes the behavior given in this table by deriving the equations for the next states of Q and QN based on the current inputs. These are:

$$Q^+ \equiv CD + QN' \tag{5.3}$$

and

$$QN^+ \equiv CD' + Q', \tag{5.4}$$

as can readily seen by inspection. When $C = 1$ and $D = 0$ (row 0 of the table), then Equations 5.3 and 5.4 reduce to $Q^+ \equiv QN'$ and $QN^+ = 1$, respectively. Once QN^+ becomes 1, then $Q^+ = QN' = 0$, and thus row 0 of the truth table is verified. If $C = 1$ and $D = 1$ (row 1 of the table), then the governing equations become $Q^+ = 1$ and $QN' \equiv Q'$. Once Q^+ becomes 1, QN is forced to become 0, verifying the second row of the table. Finally, consider the case where the circuit is not enabled, i.e., $C = 0$ (row 2 of the table). Then Equations 5.3 and 5.4 reduce to $QN^+ \equiv Q'$ and

$QN^+ = Q'$. Thus, as long as Q and QN are complementary, as they will be normally after a set operation, then they will remain in the same state. If not, whichever gets set first will stay in that state, and will force the other to be its complement.

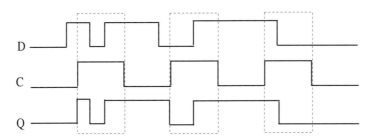

FIGURE 5.4 Example timing diagram for the D latch.

The D latch is also known as a transparent latch because it writes whatever is on D to Q, without modification, when enabled. We can see this behavior in the sample timing diagram in Figure 5.4, in which C acts like a clock, or a regular alternation between logic 0 and logic 1. When C is low, Q maintains its state, whatever that happens to be. However, when C is high, Q will mirror D. This is indicated in the dotted boxes. For example, when C first goes positive, Q is low and D is high. Therefore, Q, reflecting D, immediately also goes high. Likewise, when D drops and then returns during this clock cycle, Q reflects this. It then stays high during the off cycle of the clock because this is the maintain (no change) state for this latch. This simplicity of operation will prove useful in constructing the D flip-flop, described in the next section.

ON THE CD

System Dlatch on the CD-ROM contains cells for the input and output variables in the D latch, namely, one for each of C, D, Q, and QN. Following Figure 5.3, it would be possible to emulate this latch by first generating the outputs of the set of NANDs on the left and then the NANDs on the right (recall that an OR with all inputs complemented is equivalent to a NAND gate). However, there is no reason not to implement Equations 5.3 and 5.4 directly, and this is the method used in this system. Setting C to 1 and varying D confirms the first two rows of Table 5.2. Note that two simulation steps are needed to set the values for Q and QN correctly in both cases, in accord with the explanation above. When C is 0, Q and QN remain in the same state regardless of the value of D. This is the maintain state for the D latch. When C is 0 and Q and QN are equivalent, then an oscillation occurs. Exercise 5.8 asks you to explain why this is the case.

FLIP-FLOPS

A latch contains much of what we are looking for in a circuit that can be set and be made to hold a state, but is not ideal for the construction of sequential circuits. The problem is that the transparency of the latch with C = 1 means that any shift in the input D during that time will also affect the ouput Q. Imagine for example, driving a D latch with another D latch. The second latch will change value soon after the first latch changes, and will reflect this change. But what we require is that for all latches to respond to the prior state of the system (that is, all the previous states of the latches).

In order to achieve this synchrony, what is needed is a circuit that is not only clock-driven, but also one in which the state is set at a single instant in time, or something as close as possible to this condition. The longer the period in which a transition can take place, the more difficult it is to guarantee that the correct signal will set the device. When this time can be made sufficiently short, the circuit is termed a flip-flop rather than a latch. That is, it possibly flips its state in response to a near-instantaneous pulse rather than continuously reflecting the change in the input to the device. Here we will consider two types of flip-flop, the edge-triggered D flip-flop, and the edge-triggered J-K flip-flop.

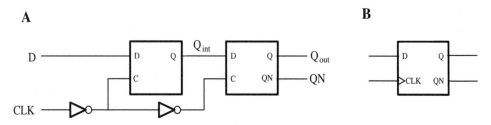

FIGURE 5.5 (A) The edge-triggered D flip-flop and (B) the logic symbol for this flip-flop.

Edge-Triggered D Flip-Flop

Figure 5.5 shows how an edge-triggered D flip-flop can be constructed from two D latches. In describing the operation of this device, it will be especially useful to bear in mind the high-level description of the D latch as a device that transparently copies the D signal onto the Q output. When the CLK signal is low, the input to C on the first latch is high; thus, during this time D is written directly to the intermediate output, Q_{int}. However, the input to C on the second latch is low, and thus

Q_{int} is not written to Q_{out}. Consider what happens when the clock signal becomes positive. At this point, the first latch becomes disabled, but the second latch becomes enabled. This means that whatever signal is present at Q_{int} is written to Q_{out}. However, because the first latch is not writing to its output, any change on the input line D will have no further effect on the final output, Q_{out}. Only the value of D right before the leading edge of the clock (that is, as it turns positive), will be written to the output.

The truth table in Table 5.3 encapsulates this behavior. Whenever the clock signal is not proceeding from low to high, the circuit is in the maintain state and the prior value of Q and QN are constant. Only during the rising edge of the clock will the input D be written to the output Q.

TABLE 5.3 The Truth Table for the D Flip-Flop

D	CLK	Q	QN
0	⌐	0	1
1	⌐	1	0
x	0	last	last
x	1	last	last

Figure 5.6 shows a typical timing diagram for this flip-flop and elucidates the operation of the circuit by also showing the internal signal Q_{int}. This signal is just a mirror of D, but only when the CLK signal is low, because the first D latch is driven by the inverted form of this input. This takes place, as before, in the areas indicated by the dotted boxes. Then, on every leading edge of the CLK, Q_{int} is transferred to Q_{out}. This is indicated by the arrows. By observing both D and Q_{out}, it is easy to see that the latter takes the value of the former on leading clock edges, and otherwise stays constant, as desired. This behavior is particularly well suited to the design of sequential circuits, as we shall see in Chapter 6.

ON THE CD

System Dflipflop on the CD-ROM contains an emulation of this device. As in Figure 5.5, two D latches are present. Each of the latches is governed by Equations 5.3 and 5.4, as can be seen by observing the truth tables and variable definitions for each cell. The first latch is driven by the inverted clock signal, and the second directly by the clock signal C. Stepping twice causes Q_{int} to register the value of

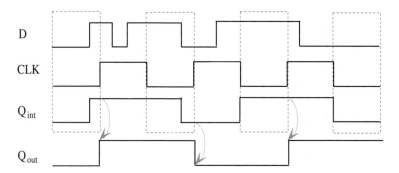

FIGURE 5.6 Example timing diagram for the D flip-flop.

the input at D1. Note that the clock C is off; thus the D signal is held at this intermediate level (C1, the complement of C, is on, but the clock for the second latch, which is identical to C, is off). Left-clicking on C will turn the clock on. This, in turn, causes Q_{int} to pass its value onto Q_{out}, as can be seen by stepping twice again. Note that Q_{int} does not change during this process. If we then set D to off, this will not affect Q_{int}. It is only after turning the clock off, and then back on again (i.e., presenting a rising clock signal to the system), that this new value of D is passed onto Q_{out}.

Edge-Triggered J-K Flip-Flop

Although the D flip-flop is the easiest to use in construction of sequential circuits, there are occasions when other flip-flops produce a more efficient implementation.

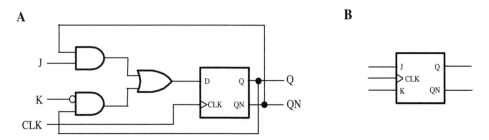

FIGURE 5.7 (A) The edge-triggered J-K flip-flop and (B) the logic symbol for this flip-flop.

One such circuit is the edge-triggered J-K flip-flop; Figure 5.7 shows one realization with the just-discussed D flip-flop as the centerpiece of the device. If the CLK signal is not experiencing a leading edge, then Q and QN are maintained in their former states because this is the only time that D is written onto Q. This accounts for the first two rows of the truth table in Table 5.4. The other rows correspond to the times in which there is a leading edge of the clock, for each of the four combinations of J and K values. In these cases

$$Q^+ \equiv QN\,J + QK' \qquad (5.5)$$

and

$$QN^+ \equiv Q'. \qquad (5.6)$$

These equations follow from the fact that D is written onto Q on a leading edge of the clock. In other words, the next state of Q is given by the SOP logic leading into the D input of this flip-flop, and QN will be the complement of Q as in the D flip-flop. When J and K are both 0, Equation 5.5 reduces to $Q^+ \equiv Q$, and therefore QN^+ also stays constant. When $J = 0$ and $K = 1$, Equation 5.5 reduces to $Q^+ \equiv 0$ and therefore $QN^+ \equiv Q' \equiv 1$. When $J = 1$ and $K = 0$, $Q^+ \equiv QN + Q$. In this case, as long as Q and QN are complementary, as they will ordinarily be, $Q^+ = 1$ (because any signal added to its complement will be 1), and therefore $QN^+ = 0$. Finally, when both J and K are 1, $Q^+ \equiv QN$, and therefore the outputs flip their state. This possibility gives the J-K flip-flop added flexibility over the D flip-flop.

TABLE 5.4 The Truth Table for the J-K Flip-Flop

J	K	CLK	Q	QN
x	x	0	last	last
x	x	1	last	last
0	0	⌐	last	last
0	1	⌐	0	1
1	0	⌐	1	0
1	1	⌐	last QN	last Q

Figure 5.8 shows a timing diagram for the J-K flip-flop. For simplicity, this diagram moves successively through each of the last four rows of Table 5.4, with each rising edge of the clock numbered accordingly. During rising edge 1, J and K are 0, so Q and QN do not change. J and K are 0 and 1 respectively during rising edge 2, so Q drops to 0 and QN rises to 1. The opposite occurs during rising edge 3. Finally, Q and QN flip values during rising edge 4.

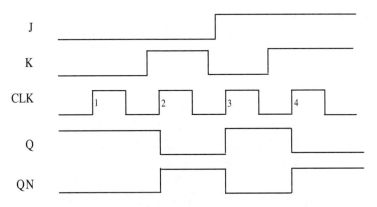

FIGURE 5.8 Example timing diagram for the J-K flip-flop.

ON THE CD

In order to implement a J-K flip-flop, it suffices to start with a D flip-flop and add the extra logic for the D input. Thus, system Jkflipflop on the CD-ROM is system Dflipflop with the addition of two cells, J and K. In addition, the input to the original flip-flop, D1, now contains the logic in Equation 5.5. It is easy to confirm that unless there is a rising edge of the clock, the system stays in the same state. Turning now to the states where there is a rising edge, we begin by setting J and K to off, and setting Q and QN in a complementary state. On the leading edge of the clock (that is, after turning cell C from off to on), Q and QN will remain in the same complementary state, assuming they are already in one. Now turn K on. Q will be set to off and QN to on when C is turned on. Likewise, Q will be set to on and Q to off when J is on and K are off, on the leading edge of the clock. Finally, with J and K set 1, Q and QN will flip states on the leading edge of the clock. This last case is not possible with the D flip-flop, making it more desirable to use the J-K flip-flop in certain circumstances.

REGISTERS

Modern computing systems consist of a hierarchy of memories, ranging from external backup storage devices, to hard disks, to random-access memories (RAMs), and finally to caches, fast RAMs located near the central processing unit (CPU). The very fastest memories are the registers—usually an integral part of the processor. It is not just speed that is at stake. What is also required is a simple mechanism to read and write data, ruling out more complex memory systems such as RAMs that require extra hardware to meet this end (see Chapter 7). Fortunately, a relatively simple extension of the flip-flop achieves this goal as well as the ability to store multiple bits of information at once. Two types are registers are now described.

Parallel-Load Registers

Figure 5.9 illustrates the parallel-load register. The register consists of four edge-triggered D flip-flops, each of which is capable of storing one bit of information (actual registers will usually have a greater number of flip-flops and therefore a greater storage capacity; for simplicity, we describe the smaller register shown here). Writing is achieved simply by setting the lines $D_0 - D_3$ to the desired values. These values will then be transferred to the lines $Q_0 - Q_3$ on a leading edge of the clock, in accordance with the operation of the D flip-flop. The values can then be read from these lines as needed.

There arc two additional mechanisms shown in the figure, which will aid in its use as a storage device. First, note that every flip-flop contains a CLR line. This line is a convenience that allows the flip-flops to be reset to 0 asynchronously, that is, regardless of the state of the clock signal. In addition, the logic for the clock signal for each of the flip-flops consists of an additional line, which is known as the LOAD that is summed with the actual clock, CLK. The purpose of LOAD is to allow control over when new memories will be written to the register. Generally speaking, we do not wish to write on every clock cycle, but only when the register requires updating.

When the LOAD is 0, the output of the OR gate is always 1, and thus no writing can take place (recall that each flip-flop can only be reset on a rising edge of the clock, and if the CLK signal is constant, there are no such edges). However, when LOAD goes to 1, then the output of the OR gate will be equivalent to the CLK, and writing is permitted. In summary, a parallel-load register consists primarily of a single flip-flop for each bit that needs to be stored, and some additional mechanism to prevent loading of the register when writing is suspended. In Chapter 7, we will see how such registers function in the context of a larger computational system.

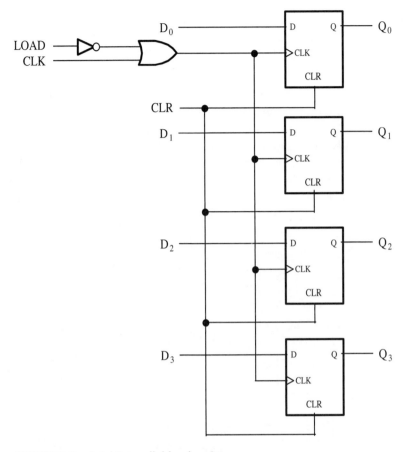

FIGURE 5.9 A 4-bit parallel-load register.

ON THE CD

System register on the CD-ROM contains a single register, as shown in Figure 5.9. This is the same as system Dflipflop except for the addition of the LOAD and CLR lines. In the default configuration, the system is poised to load a 1 bit on the D line. Stepping through the simulation writes this bit onto Q_{int}, and then after making the CLK active, this will be written to Q_{out}, because we have provided an off-to-on transition of the clock. So far, the system behaves identically to the D flip-flop. Now turn D1 off to load this into the flip-flop, but this time turn off LOAD first. Notice that no change takes place after we provided a leading edge of the clock because the load is disabled. Clicking on C reveals it to be the sum of the complement of LOAD and the CLK, as shown in Figure 5.9. Now click on CLR and advance the system. This sets Q_{int} to 0 (and Q_{out} to 1) regardless of the state of the CLK or LOAD signals. Examination of Q_{int} shows that this is accomplished by multiplying the complement of the CLR signal with the normal function for Q_{out} in this flip-flop.

Shift Registers

It is also possible to load the bits into the register one by one, which will prove convenient if they are provided serially on a single line rather than in parallel as in Figure 5.9. The shift register in Figure 5.10(A), so-called because the bits are sequentially shifted from one flip-flop to the next, accomplishes this task. On each rising edge of each clock cycle, the SERIAL IN line will load Q_0. However, at the same time, each flip-flop will be set to the output from the preceding flip-flop. As the bits enter the SERIAL line, they will be successively shifted into the flip-flops. For example, a high bit, or 1 on this line will set Q_0 to 1 at the first up-tick of the clock, Q_1 to 1 at the second, Q_2 to 1 on the third, and finally Q_3 to 1 on the fourth.

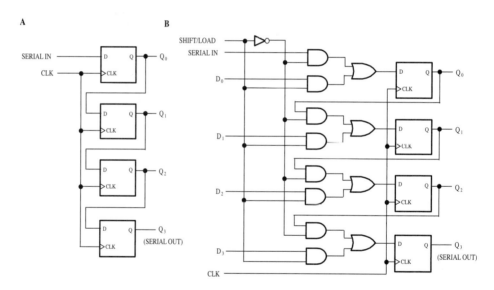

FIGURE 5.10 (A) A 4-bit shift register, serial load, and (B) a combination serial/parallel load shift register.

It is easy to see that Q3 provides a serial means of reading the registers. As the bits get shifted down, they will leave Q3 in the same order in which they entered, four clock cycles later. However, the circuit in Figure 5.10(A) also provides a means of reading the register contents directly by accessing the flip-flops outputs, Q_0 through Q_3, in parallel.

For this reason, this circuit is also known as a serial-in parallel-out register. The choice of which way to read the registers will depend on the how bits will be processed downstream of the register: if they are required in parallel, then the flip-flops will be accessed directly; otherwise they will be read one at a time from the SERIAL OUT line.

System shiftreg1 on the CD-ROM contains an emulation of the shift register in Figure 5.10(A). Here we have simplified the circuit in the figure, without destroying its essential character, by producing the behavior of the Q's without implementing each D flip-flop. This is possible because the maintain state can be simply implemented in LATTICE by defining a variable that corresponds to the cell itself.[2] In addition, the flip-flops potentially change state when the clock is 1, not on a rising edge, as in the D latch. The entire truth function for each Q_i is (Q_i * CLK') + (input * CLK) where the input is derived from the actual input to the circuit in the case of Q_0 and is Q_{i-1} otherwise. Thus, if the clock is off, the system stays in its previous state, and if on, it reads the input. Successively stepping through the simulation illustrates the full progression of events. A bit stream to the left is steadily written to the Q_0. These are then shifted down into the Q below, until they reach Q_3, and then are placed on the serial out line. In this simulation, each step represents one rising edge of the clock in the full circuit in Figure 5.10(A). The serial in and out lines are constructed to propagate their contents from left to right.

Figure 5.10(B) shows a register with even greater flexibility. As with the prior circuit, the register outputs can be read in serial (from Q_3 alone) or in parallel (from Q_0 to Q_3 simultaneously). In addition, the data can be either loaded in parallel, as in Figure 5.9, or in serial, as in Figure 5.10(A). This is achieved by attaching the logic for a 2-input 1-bit multiplexer to each of the flip-flop inputs, where the SHIFT/LOAD line serves as the select bit. When this line is set to logic 0, the SHIFT condition, the first (topmost) AND gate in each of these subcircuits is enabled, and the circuit is virtually equivalent to Figure 5.10(A). When this line is set to logic 1, then the second AND gate is selected and the circuit becomes virtually equivalent to Figure 5.9 in that the D lines can be loaded in parallel. For simplicity, not shown in either Figures 5.10(A) or 5.10(B) is a mechanism that only allows register setting during desired time intervals. Such mechanisms are usually present and are similar to that discussed in the prior section in that they provide a gated clock signal rather than a raw clock signal.

System shiftreg2 on the CD-ROM works very much like shiftreg1 except that the input to each flip-flop is replaced by an SOP expression corresponding to the logic driving this input on each circuit. This can be seen by right-clicking on any of the Qs and observing the truth table (you will need to expand the truth table window by clicking on the right arrow on the expansion bar between this window and the animation window). The previous simple input has been replaced with (S/L' *

input) + (S/L * D$_i$), where the input is the Q from above (or the serial input in the case of the first flip-flop), S/L is the shift/load signal, and D$_i$ is the parallel load signal for each flip-flop. In the default configuration, shift/load is inactive, and thus the circuit acts as a shift register. This can be seen by advancing the simulation; it works just like shiftreg1. Now click on shift/load. The circuit will load the Ds onto the Qs in parallel, in this case, setting Q0 and Q2 active.

SUMMARY

If we are to create a device of any complexity, it must be capable of holding its state and responding to inputs as a function of that state. In this chapter, we introduced a number of ways to do this. First we discussed the SR latch. It is possible to put this latch into one of two states and to keep the latch in its current state, with the settings of the S and R input lines. The D latch is conceptually simpler than the SR latch in the sense that there is only one input line. The latch is set to this state whenever it is enabled. The D latch also sets the stage for the D flip-flop, which is enabled not by a constant signal, but by the rising edge of the clock. This type of device will prove useful in subsequent circuits in which we wish to narrow the window in which the device can be set. The J-K flip-flop achieves even more flexibility, albeit with more gates, in the sense that the proper set of inputs can cause the state to be inverted, regardless of the current state. To store more than one bit, we can put a number of flip-flops together to form what is known as a register. Registers come in many varieties, the simplest of which is the parallel load, in which each flip-flop can be set separately. It is also possible to load a register in a series, by having the output of one register feed the input of the next; this is known as a shift register. Finally, one can combine the two loading methods to form a single register, and have a selection line determine which method to use. Flip-flops and registers will form the core of many of the circuits to be developed in subsequent chapters.

EXERCISES

5.1 The table below shows the signals on the S and R lines of an SR latch. Fill in the rest of the outputs Q and QN, assuming the following convention. The outputs in a given column are a function of the inputs in the same column, that is, it is implicitly assumed that these outputs reflect the next state of the machine. For example, the first column in the table shows the result of the set operation.

5.2 Fill in the table below, using the prior convention, and assuming
   (a) The device in question is a D latch.

Elements of Sequential Design   **135**

S	1	0	1	0	0	0	0
R	0	0	1	0	0	1	0
Q	1						
QN	0						

(b) The device in questions is a D flip-flop (assume that the value before the leading-edge is the one just to the left of this event).

D	1	1	0	0	1	0	0	1	1	1	0	1
CLK	0	0	0	1	1	1	0	0	0	1	1	1
Q	0	0	0									
QN	1	1	1									

5.3 Figure 5.11 shows a D flip-flop, however, unlike the one in Figure 5.5, this writes on the falling edge of the clock. Prove that it works correctly by filling in the timing table below, that is, show that D gets written onto Q only when the clock goes from 1 to 0 (use a similar assumption to that in Exercise 5.2(b).

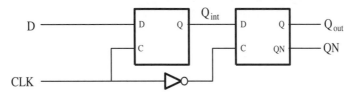

FIGURE 5.11 The falling edge-triggered D flip-flop.

D	1	1	0	0	1	0	0	1	0	1	0	1
CLK	1	1	1	0	0	0	1	1	1	0	0	0
Q_{int}	1	1	1									
Q	1	1	1									
QN	0	0	0									

136   Digital Design: From Gates to Intelligent Machines

5.4 Fill in the following table for the shift register in Figure 5.10(A), assuming that each bit arrives on successive leading clock edges.

SERIAL IN	0	1	1	0	1	0	1	1	0
Q_0	0								
Q_1	1								
Q_2	0								
Q_{33}	1								

5.5 Fill in the following table given the following inputs to the shift register in Figure 5.10(B).

5.6 Figure 5.12(A) shows an alternative design for the SR latch. Show that it has the same truth table as the design in Figure 5.1(A), with the exception of the last (and least important) row.

5.7 Figure 5.12(B) shows the design for a latch. Produce the truth table for this latch. Why is it called a toggle latch?

SERIAL IN	1	0	1	1	0	0	1	1	0
SHIFT/LOAD	0	0	1	1	1	0	1	1	0
D_0	1	1	1	0	1	1	0	1	0
D_1	0	1	0	1	0	0	0	1	1
D_2	0	1	1	0	0	0	1	1	0
D_3	1	1	1	0	1	1	0	0	1
Q_0	1								
Q_1	0								
Q_2	1								
Q_3	0								

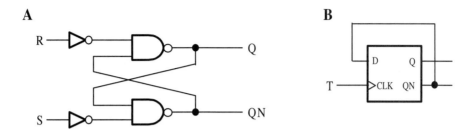

FIGURE 5.12 (A) An alternative design for the SR latch and (B) the design for a new latch.

LATTICE Exercises

5.8 Use system Dlatch to explain why this latch can enter into an oscillating state.

5.9 Implement the alternative SR latch in Figure 5.12(A) in LATTICE.

5.10 Implement the toggle latch in Figure 5.12(B) in LATTICE.

ENDNOTES

1. In this and other derivations in this section, we make use of the logical laws that $0 + X = X$ and $1 + X = 1$.
2. Every action must "bottom out." Here we have a program that is simulating the hold state of a latch, and therefore somewhere there must be a way of holding a memory state. In this case, the program variables (LATTICE is written in Java) correspond to locations in random access memory (RAM), the contents of which are maintained as long as the computer is on and they are not overwritten (see Chapter 7).

6 Sequential Machines

In This Chapter

- Introduction
- Finite State Machines
- Mealy and Moore Machines
- Sequential Machine Analysis
- Sequential Machine Synthesis
- Designing with J-K Flip-Flops
- Summary
- Exercises

INTRODUCTION

To see why we need sequential machines, let us return to an earlier example from Chapter 3. In that chapter, we considered the case where every other square of the Karnaugh map was filled in a checkerboard pattern. We also discussed how this couldn't effectively be minimized with NAND gates although a large XOR gate would suffice. This is fine as far as it goes, but what happens if the variable count is greater than four? We could, of course, build a very large XOR gate or create a multilevel XOR function once fan-in restrictions are hit. When the number of input variables hits 128, 256, or even 1024, however, this is impractical, as it would entail a very large circuit. Moreover, what if we didn't know the number of input variables in advance? A fixed combinational circuit can only respond to an input of constant length.

The key to the solution, as with many problems in computer science (both hardware and software), is to trade space for time. Rather than trying to process the inputs in parallel, we will process the inputs one at a time. This will take longer than the parallel (and combinational) process, but will yield a circuit that is both more compact and more flexible than that provided by previous methods. To accomplish this, we need machines with internal states, which, as you may recall, was the purpose of introducing latches and flip-flops in the earlier chapter. In effect, the structure of these machines will allow past inputs to influence current outputs via the states that are realized as the result of the processing of these inputs. In the next section, we introduce a formalism, the finite state machine, that provides a formal structure for the design of such machines.

FINITE STATE MACHINES

We begin by recasting our original description of a digital circuit as a device that transforms inputs to outputs to one that produces one or more outputs in the presence of an input string. By a string, we simply mean a sequence of characters, such as "abaca" or "10111001." By an output we simply mean that it responds with a "yea" (or 1 or active) when it sees strings of a certain type and a "nay" (or 0 or inactive) otherwise. For example, our characteristic example discussed above is an odd parity machine, that is, it responds positively if the count of the 1s is odd and 0 otherwise. To "0110111" (five total 1s) and "1011" (three total 1s) it would signal an active output and for "1010" (two total 1s) and "111111" (six total 1s) it would not. Note that such a machine does not place any limits on the length of the string; it could have 4, 16, or 1024 elements.

A finite state machine is an abstract representation of the device that accomplishes this task. It consists of three components. First, there are the states themselves. Then there are the inputs, which effect the transitions between the states. In the case of DFSMs (deterministic finite state machines), the kinds of machines under consideration here, there will be a unique transition for each possible input character. The defining characteristic of the DFSM is that the next state of the machine is completely determined by the current state and the current input. Thus, if the input strings were composed of the letters a, b, c, and d, then each state will be characterized by a total of four transitions, one for each of the characters. Finally, there is the output of the machine, signaling that the string is or is not of the correct form, or perhaps other intermediate level cues as to the state of the device.

Figure 6.1 shows an example of a DFSM that recognizes odd parity, our characteristic example. It consists of two states representing even and odd parity. The even state means that an even number of 1s have been received so far (this is also the starting state of the machine because at the start zero 1s have been received and zero is even). Likewise, the odd state implies that an odd number of 1s have been processed. The arcs connecting the states indicate what to do on each successive input character. In this case, there are two possible characters, 0 and 1, and thus there are two arcs from each state. In the even state, receiving a 0 means staying in the same state, and a 1 implies a transition to the odd state. In the odd state, receiving a 0 also means staying put, and a 1 implies a transition to the even state. Additionally, in the odd state, we generate a recognition signal to indicate that an odd parity string has been received so far, represented in the figure by the shading of this state.

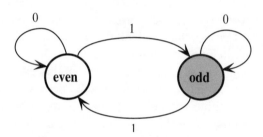

FIGURE 6.1 A DFSM that recognizes odd-parity strings.

It is easy to see that this machine will only generate an output when an odd number of 1s are processed. For every 1 that is received, the state changes. If one starts in the even state, and an even number of 1s are received, then an even number of state changes will be effected, and the machine will continue to be in the even state. Conversely, if there are an odd number of 1s, then an odd number of state changes will take place, and the machine will end up in the odd state, and will signal odd parity.

ON THE CD

System parity in the Chapter6 folder on the CD-ROM is our first illustrative example of a state machine, other than the music machine from Chapter 1 that was presented without explanation. Unlike a truth table machine, these machines are driven by state tables. Such tables are formally equivalent to the kind of state diagram seen in Figure 6.1, the chief difference being each of the possible transitions are represented at the top of the table, and the states are located in a single column on the left. In addition, the states are color coded for added visibility in LATTICE.

Finally, LATTICE allows each state to generate an output. In this case, the state flashes when odd parity is present. It is also possible to associate each state with a music file or a picture.

In this example, when the parity machine is in the even state, and I0 = 0, then the machine stays in the even state (the definition for I0 can be seen by left-clicking on the cell at the top of the column labeled "I0"; it corresponds to the presence of a blue bit to the left of the parity cell). When there is a blue bit to the left, that is, I0 = 1, it goes into the odd state. Likewise, the odd state remains static when I0 = 0, and progresses to the even state when I1 = 1. The bitstream machine to the left just advances the bits to the left for input to the parity machine.

Thus, the parity machine absorbs the bits that come in, with a blue bit affecting a transition and a yellow bit having no effect on the state. Advance the simulation to confirm that the machine only flashes when in the odd parity state. Note that when the bits run out, you can still click on a bit anywhere in the bitstream to send another blue cell on its way toward the detector, and it will flip the state of the parity machine. This confirms our earlier claim that unlike a combinational circuit for the detection of parity, this sequential machine will operate on an unspecified and potentially infinite number of bits.

We now consider a somewhat more complex example, in which the task is to recognize input strings of the form "abac." There is a simple trick[1] in constructing a finite state machine whenever the task is to recognize a particular sequence, which we illustrate with this example in Figure 6.2(A). Starting with an initial state, which we will call INIT, you need only label each successive state with the partial sequence leading up to that state. Thus, in this example, the states are "a," "ab," "aba," and finally the recognition state, "abac." The symbols on the arcs correspond to the letters that are needed to reach the given state. It is easy to see that starting with the INIT state, the sequence "abac" leads directly to the recognition state.

Once this skeleton is formed, it then remains to fill in what happens when other letters are received; this is shown in Figure 6.2(B). To accomplish this in the case of sequence recognition, we use the following rule: If the current character, possibly in conjunction with any previous characters, is equivalent to a partial starting subsequence of the desired string, then go to the state corresponding to this subsequence; otherwise, go back to INIT. For example, when the current state is "aba" and an "a" is received, we go back to state "a" rather than INIT ("a" is a partial starting subsequence of "abac"). This allows proper recognition if the sequence is "ababac," for example, and would not happen if we went back to INIT (see Exercise 6.2). With this procedure, we allow the system to recognize that the last four characters in the sequence represent the specified target.

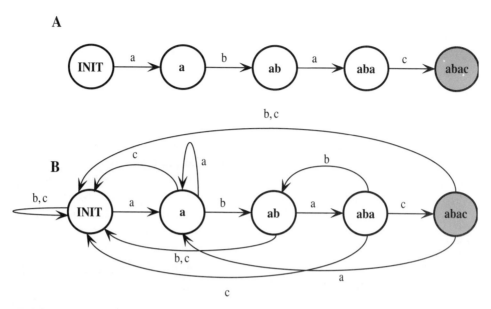

FIGURE 6.2 (A) A DFSM skeleton formed by including only those transitions on the recognition path, and (B) the full machine with all transitions in place.

To take another example, if we are in state "aba" and receive a "b," we go back to state "ab" rather than INIT. In this case, the last two letters are "ab," thus forming a potential start for the full desired string. For example, if the string were "abababac," we also want to consider this as acceptable input string (the last four characters match the target). This can only be done if, on receiving the second "b," we recognize that we have a starting sequence "ab," which potentially forms the start of a new target sequence. Please confirm for yourself that all the transitions in the table conform to this rule of partial matches.

State diagrams such as those in Figures 6.1 and 6.2 are useful because they allow us to visually grasp the nature of the progressions in the machine with a minimum of effort. The alternative, and the one used in LATTICE, is a state table. With a state table, it is somewhat more difficult to see what is going on, although the tabular form does have the advantage that it is more easily converted into the governing equations for the sequential circuit, as we will see in subsequent sections.

ON THE CD

System sequence1 on the CD-ROM contains the finite state machine recognizing the bit string "abac." After loading this system, you can see the sequence lined up and ready for input to the sequence machine with turquoise representing "a," light blue "b," and dark blue "c" (the sequence is read from right to left because it will be proceeding rightward). You can also observe that the state table corresponds to the DFSM given in Figure 6.2(B). In this table, I0 = 1 represents the presence of an "a" to the left of the sequence machine, I1 = 1 represents the presence of a "b," and I2 = 1 the presence of a "c." The only transitions of significance for this table are the ones where one of these variables is 1 and all others are 0; these are the cases that correspond to the transitions in the DFSM (later in this chapter we will do a binary encoding of transitions and thereby produce a more compact table). Successively stepping through the simulation affects the transitions from one state to the next, until the gold, or recognition state, is reached after the last character is processed. Now open the pattern sequ-pat2. This contains the string "ababac." After processing "aba," the sequence machine moves to the state corresponding to this substring. The interesting case is next. On receiving another "b," proceed not back to INIT, but to the state corresponding to the substring "ab" (dark blue) because this is an initial subsequence of the correct string. And in fact, when "ac" is received after this, the machine correctly moves to the recognition state. It effectively ignores the first "ab" in the input string "ababac" and realizes that the last four characters are correct.

MEALY AND MOORE MACHINES

We will construct two classes of machines: Mealy and Moore. Figure 6.3 illustrates the difference between these machines and provides a high-level description of a sequential machine. The next-state logic contains the circuitry that determines how inputs and prior states affect future states. The state memory, typically consisting of a set of flip-flops, holds the current state. The output logic determines how output values are computed. In the case of a Moore machine, these outputs are strictly a function of the current state. In the case of a Mealy machine, the output logic is also influenced by the current inputs.

The event sequence for these machines is as follows. At any given time, the machine is in a given state, as encoded by the state memory. On the rising edge of the clock (for rising edge D flip-flops), or another clock-based trigger for other types of flip-flops, the next-state logic combines the current input information with the prior state to place the machine in a new state. If the device is a Moore machine, this state will determine the values of one or more outputs. If it is a Mealy machine, then the input at the time of the rising edge will also determine the outputs. These ideas will be further illustrated in the next two sections.

LATTICE implicitly assumes that you are creating a Moore machine because an output (a flash, a picture, or a piece of music) is generated when a state is entered, regardless of how that state was reached. However, you can simulate a Mealy

Sequential Machines 145

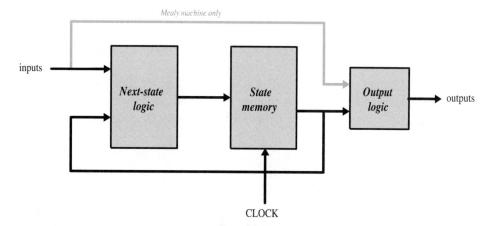

FIGURE 6.3 A block diagram illustrating the basic operation of a sequential machine and the difference between Mealy and Moore machines. The direct influence of inputs on output logic is present only in a Mealy machine (gray line).

ON THE CD

machine by looking at current inputs and states and having a machine that responds to both. System soda in the Chapter6 folder on the CD-ROM illustrates such a system. This machine will generate a soda if two nickels are entered and the button is pushed (left click on the "nickel in" machine and advance the simulation twice, and then click on the button machine and click on advance). Also, if three nickels are entered without hitting the button, then it will return a nickel. In both cases, outputs are generated by looking at the state of the inputs and the state of the machine itself (the soda machine).

SEQUENTIAL MACHINE ANALYSIS

You will most often be called upon to design a sequential machine rather than analyze an existing machine. Nevertheless, analysis serves as a useful introduction to what can sometimes appear to be a formidable process. There are six steps in analyzing a sequential machine constructed from D flip-flops:

1. Derive the transition equations from the next state logic.
2. Derive the output equation from the output logic.
3. Construct a transition/output table from the equations produced in (1) and (2).

4. Construct a state/output table by labeling the states in (3).
5. Derive a DFSM from the table in (4).
6. State in ordinary language what the machine is doing.

We now apply this process, step by step, to the sequential machine diagrammed in Figure 6.4. Although this circuit may appear complex, adherence to the above steps reveals its behavior in relatively short order.

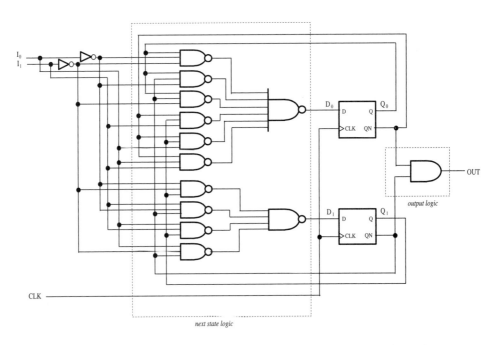

FIGURE 6.4 A sequential circuit consisting of two edge-triggered D flip-flops, the next-state logic for the flip-flops, a clock signal feeding both, and output logic driving the output signal.

1. Derivation of the transition equations

Here we take advantage of the fact that although the circuit taken as a whole is sequential, the guts of the machine are combinational. As you can see, the inputs to each of the two flip-flops are nothing more than two-level NAND circuits, which, as you should recall, implement SOP expressions. Although it may be difficult to unravel the spaghetti-like nature of the connections, with a little care it can easily be seen that the transition equations for each of the flip-flops are:

$$Q_0^+ = Q_0I_0'I_1' + Q_0Q_1'I_0' + Q_0Q_1'I_1' + Q_0'Q_1I_1 + Q_0'Q_1I_0 + Q_0'I_0I_1, \text{ and} \quad (6.1)$$

$$Q_1^+ = Q_1I_0'I_1' + Q_1'I_0'I_1 + Q_1I_0I_1 + Q_1'I_0I_1'. \quad (6.2)$$

In deriving these equations, we use the fact that the D flip-flop transparently writes the D input to the flip-flip onto the Q output at each rising edge of the clock. Thus, the next states of the machine, the Q^+s, can be obtained directly from the logic to the Ds, as in Equations 6.1 and 6.2.

2. Derivation of the output equation

The final output is a function of the flip-flop outputs only (thus, this is a Moore machine) and is easily seen to be

$$\text{OUT} = Q_0'Q_1', \quad (6.3)$$

given that the QNs are the complement of the Qs for each flip-flop.

3. Construction of the transition/output table

In order to understand what the machine does, we wish to know how the inputs influence the transitions from one state to the next, and how the states (and possibly the inputs in the case of a Mealy machine) influence the output. The transition/output table in Table 6.1, which details the next states, Q_0^+ and Q_1^+ as a function of the prior states Q_0 and Q_1 and the current inputs, is a compact representation of this information. There is a laborious, albeit straightforward process that allows us to fill in the values of the table. Each cell is computed directly from the transition Equations 6.1 and 6.2 for each of the two future states. For example, suppose we are attempting to compute the value of the shaded cell. Equation 6.1 tells us that when Q_0, Q_1, I_0, and I_1 are 0, 1, 1, and 0 respectively, that the value of Q_0^+ will be 1. Likewise, when these are the current values, Equation 6.2 tells us that the value of $Q1^+$ will be 0. Therefore, the content of this cell is 10, representing the fact that the next state variables will take these values if the current state $Q_0 = 0$ and $Q_1 = 1$ and the input is 10. The output column is computed directly from the output Equation 6.3. In this case, the output is a function of the state variables only; thus, the inputs have no bearing on these values.

TABLE 6.1 The Transition/Output Table for the Machine in Figure 6.4

Q_0Q_1	I_0I_1				OUT
	00	01	10	11	
00	00	01	01	10	1
01	01	10	10	11	0
10	10	11	11	00	0
11	11	00	00	01	0
	$Q_0^+Q_1^+$				

4. Construction of the state/output table

In principle, we have completed our analysis. Table 6.1 contains a complete description of what the sequential circuit will do in every circumstance. However, the purpose of analysis is to allow for human interpretation, and this table will be opaque to most observers. Therefore, two extra steps of clarification are now typically carried out. In the first, we replace each state combination with a unique letter. In this case, there are two state variables, and thus $2^2 = 4$ possible states. Each is labeled successively with A through D, both in the left column showing the original state, and in the 4×4 grid in the center of table, which shows the next state of the machine as a function of the inputs and the prior state. The result is shown in Table 6.2, which is just Table 6.1 with A, B, C, and D replacing 00, 01, 10, and 11 respectively.

TABLE 6.2 The State/Output Table Corresponding to Table 6.1

Q_0Q_1	I_0I_1				OUT
	00	01	10	11	
A	A	B	B	C	1
B	B	C	C	D	0
C	C	D	D	A	0
D	D	A	A	B	0
	$Q_0^+Q_1^+$				

5. Derivation of the finite state machine

At this point, it is possible to transform Table 6.2 directly into a finite state machine. This is diagrammed in Figure 6.5. The machine contains a circle for each of the states, and the transitions are those given in the table. As usual, each state contains a unique transition for each of the possible inputs, four in this case (some transitions have two labels indicating that two inputs lead to the same state, and these are counted twice). The "A" state is also shaded to indicate that an output is generated when the machine is in this state.

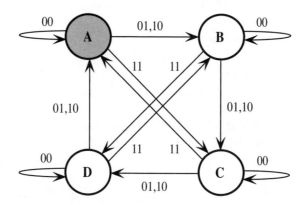

FIGURE 6.5 The finite state machine for the circuit in Figure 6.4.

6. Ordinary language description

It is now possible to state what the machine does. In the machine in Figure 6.5, the critical clue is the symmetry between the states of the machine. Whenever the input is 00, the machine stays in the same state. Whenever a 01 or 10 is received, then the machine jumps to the next state. And finally, whenever a 11 is received, the machine jumps two states. It is therefore clear that some sort of counting is taking place. If we add up the number of bits in the string, the solution immediately becomes apparent: the machine always jumps n states, where n is the sum of the bits that are 1 in the input.

In addition, because a positive output is generated in the initial state, the machine is counting to four and starting over again. In other words, the machine is a "mod 4" counter, in which the count is advanced by the number of bits that are 1 in the input string, and the states A through D represent 0 through 3, respectively, as the result of the mod operation. For example, if we start in the initial state "A," and the sequence of inputs received successively at each uptick of the clock is 10, 11, 11, 01, 10, 10, then the successive states are "B," "D," "B," "C," "D," and "A." The machine ends up in the state generating an output, and if our analysis is correct, the total number of bits received divided by four should be an integer. It is. (The total bits over the entire sequence is eight.)

ON THE CD

System counter on the CD-ROM reproduces the behavior of the finite state machine in Figure 6.5. Right-clicking on the counter machine reveals that it is an exact reproduction of Table 6.2 and therefore a faithful representation of this machine. The machine to the left of this, input, simply advances these bits to the left to be read by the counter. The counter variables are defined such that $I0 = 1$ when a bit at the left is present and $I1 = 1$ when a bit diagonally below is present. The example in the default configuration is the one just given. Advancing the machine shows that it proceeds to the desired states and returns to the initial output state after the inputs have been processed, as indicated by the flash representing the fact that the initial state A is an output state. Try playing with the inputs to observe further the behavior of this counter.

Although there is no algorithm for turning a DFSM into a natural language description, careful analysis of the transitions (as in this case), helps form the desired description. Another important clue can be found in the states (or transitions) that generate outputs. Between these two sets of indicators, it is usually possible to summarize succinctly the machine's behavior, if such a description exists.

SEQUENTIAL MACHINE SYNTHESIS

Sequential machine synthesis, as already mentioned, is quite a bit more complex that analysis. Nevertheless, by carefully following the list of steps given next, which mirror in reverse the steps in the analysis process, you should not experience undue difficulty. Later, as the process becomes more familiar, you can dispense with the recipe and proceed more directly to the solution.

General Method

1. Finite state diagram construction

One normally begins with a word description of the problem to be solved. However, this is usually too ill-defined to transform directly into a design. The most systematic way to proceed is to convert a possibly vague collection of notions into a DFSM. This has the added benefit of clarifying what the machine should do in one's own mind before wasting too much time on the details of the design process.

2. State/output table construction

The conversion of the state diagram to a state table is straightforward. Each state in the DFSM corresponds to a state entry in the table, and each transition from each of the states corresponds to an entry in the input section of the table. For example, if there are four states and three transitions from each state then the state output table will contain four rows and three additional entries for each of the states indicating where the states progress to for each of the inputs. In the case of a Moore machine, there will be an additional column indicating the output for each of the states. In the case of the Mealy machine, each cell in the table corresponding to a transition will also contain an output (this follows from the fact that the outputs are functions of the current input as well as the state).

3. State-variable assignment

The states at this point are in symbolic form. These need to be converted into a binary representation in order to be used in a digital circuit (if we were constructing a sequential machine in software, we could keep the state designations in symbolic form, but this option is not available for hardware realizations). This is accomplished by assigning a unique binary designation to each state. If there are n states, then we need m binary digits, where 2^m is the smallest power of two greater than n. For example, if there were five states, then m would be 3 (if m were 2, then we could only represent four possible states).

Then, states are taken in sequential order and given the appropriate binary designation. For example, the states A, B, C, D, and E would be given the binary assignments 000, 001, 010, 011, and 100 respectively. Note that not all possible binary combinations need to be used. In this case, 101, 110, and 111 are missing. The consequence of this will be discussed in the context of the second design example below. Note also that the assignments are somewhat arbitrary. For example, there is no reason that state A could not be 010, B could not be 011, etc., as long as each state has a unique binary identifier. Under certain circumstances, some labelings may result in a more compact circuit than others, but for the purposes of this book, we will ignore this nuance.

4. Substitution of state assignments into the state/output table

The assignments created in the prior stage are then substituted into the table constructed in state 2. In effect, a table that looks like Table 6.2 ends up looking like Table 6.1.

5. Construct an excitation table

If we are using D flip-flops, then this is the easiest step (in fact, this is the easiest task in the book). A D flip-flop transparently copies the input on the D line to the Q output. This means, as we have seen, that to achieve a given Q at the next time step, it suffices to set D to that value on the prior time step. The net implication of these facts is that the excitation table, which describes the inputs to the D lines of the flip-flops, is identical to the state table. In the final example in this chapter, we describe what needs to be done if D flip-flops are not used.

6. Derive excitation equations from this table

The goal in this step is to derive the next-state logic for each flip-flop from this data. Each column in each of the cells in the excitation table represents the input to a single flip-flop for a given state/input transition. For each of the flip-flops, this data is transferred to a Karnaugh map, so as to produce a minimized form of the next-state logic. The excitation equations will be the collection of (typically SOP) functions that result from the minimization for each of the maps. One complication arises in conjunction with cells on the Karnaugh map that are neither 1 nor 0. This is discussed in the context of the second design example below.

7. Derive output equations from the transition/output table

The derivation of the output equations follows a similar procedure to the derivation of the excitation equations. If the device is a Mealy machine, then the outputs will be a function of both inputs and states and the Karnaugh maps will be the same size as in the previous step. If, however, the device is a Moore machine, then the outputs will be a function of the states only, and the Karnaugh maps for the outputs will be accordingly reduced in size.

8. Draw a circuit diagram that realizes the excitation and output

Once the excitation and output equations are obtained, the realization of these equations can be achieved with the methods that we have been using throughout the course of the book. For example, if these are in SOP form, as they typically will be, then a two-level circuit consisting of NAND gates throughout will suffice.

9. Verification

There are two ways to verify the correctness of your design. The first is to confirm that the circuit realizes the finite state diagram by systematically confirming that each state transition is realized in accord with this original description. This is certainly a good idea, but a better, albeit more difficult method, involves verifying that a typical input sequence performs as expected. This method is superior in that it does not depend on the state diagram, which may itself be flawed. The difference may be likened to the difference between debugging by verifying that a computer program is consistent with a flow chart and running the program directly. The fact that the system responds to a characteristic series of inputs gives one confidence that the design is in accord with one's original objectives. It is, of course, necessary that the inputs are sufficiently rich to cover all the characteristic cases if the verification is to be meaningful.

The following sections apply these steps to a number of characteristic design examples.

The Parity Example

As the first example of this process, we revisit the parity problem established at the start of this section and detail the steps needed to construct a sequential machine with D flip-flops that solves this problem:

1. State diagram construction

In this case, we already have our starting point in the synthesis process, the DFSM shown in Figure 6.1.

2. Construct of the state/output table

The finite state diagram has two states and one input. Therefore, we require one state variable Q_0 ($\log_2 2 = 1$), and the state table will have two next states for each of these, corresponding to each of the inputs $I = 0$ and $I = 1$. This is shown in Table 6.3. The table also indicates that the odd state generates a high output (this is a Moore machine).

TABLE 6.3 The State/Output Table Corresponding to Figure 6.1

Q_0	0	1	OUT
even	even	odd	0
odd	odd	even	1
	Q_0^+		

Above the 0 and 1 columns: I

3. State variable assignment

There are only two states so it is natural to assign 0 to the even state and 1 to the odd state (a symmetrical and functionally equivalent circuit would be produced if we used the opposite assignments).

4. Transition/output table construction

These state assignments are substituted into Table 6.3 to produce Table 6.4.

TABLE 6.4 The Transition/Output Table Corresponding to Table 6.3

Q_0	0	1	OUT
0	0	1	0
1	1	0	1
	Q_0^+		

Above the 0 and 1 columns: I

5. Excitation table construction

Because we are using D flip-flops, which transparently write the D input onto the Q output on the rising clock edge, the logic for the Q⁺s and Ds will be identical. Thus, we can simply substitute D_0 for Q_0^+ in Table 6.4 to produce Table 6.5. Naturally, in the future when performing sequential synthesis with D flip-flops, we will skip step 4 and move to step 5 directly.

TABLE 6.5 The Excitation/Output Table Corresponding to Table 6.4

	I		
Q_0	0	1	OUT
0	0	1	0
1	1	0	1
	D_0		

6. Excitation equation derivation

As Table 6.5 shows, D_0 (the input to the single flip-flop) is a function of two variables, Q_0 and I. Normally we would minimize the equations that characterize the D inputs, but here no minimization is possible; each prime implicant is a single minterm only. Thus the excitation equation is:

$$D_0 = Q_0I' + Q_0'I. \tag{6.4}$$

If we wanted to, we could simplify Equation 6.4 by using a single XOR operator (see Exercise 6.3).

7. Likewise, there is no need to minimize the output logic, as it is simply

$$OUT = Q_0. \tag{6.5}$$

8. Circuit realization

The circuit diagram in Figure 6.6 implements both the next state logic as characterized by Equation 6.4 and output logic given by Equation 6.5. The former is implemented with a two-level NAND circuit and the latter by connecting the output of the flip-flop to the output.

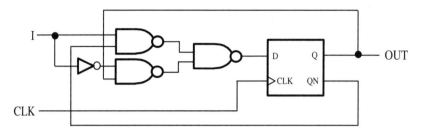

FIGURE 6.6 The sequential circuit that implements odd parity detection.

9. Verification

To verify the correctness of our design, let us subject it to the characteristic input string "01100111010." In order to do this, we begin by filling in Table 6.6 with all the values in dark; the gray values are the ones yet to be filled in. Each column in this table represents a single uptick of the clock. In the initial column, we assume that the flip-flop is in the Q = 0 state. In the other columns, the value of Q is computed from the value of Q in the previous column and the I in the current column, in accord with the next state logic given in the design in Figure 6.6. The circuit diagram also shows that OUT is equivalent to Q. As an example of the calculations of the gray values, let us compute Q and OUT for the column after the initial column. In this case, the previous value of Q was 0 and the current I is 1. Given that $Q^+ = D = Q'I + QI'$ and $Q^+ = 0' \cdot 1 + 0 \cdot 1' = 1$, OUT will therefore also be 1. The other unknown values are similarly formed.

TABLE 6.6 Verifying the Design with a Typical Input Sequence

I	0	1	1	0	0	1	1	1	0	1	0
Q	0	1	0	0	0	1	0	1	1	0	0
OUT	0	1	0	0	0	1	0	1	1	0	0

We can now check to see that the design meets the original specifications. We desire that OUT is 1 if an odd number of 1s have been received up to (but not including) the current time step and 0 otherwise. It is easily seen that this is the case at every clock cycle and thus the design can be considered correct, given the assumption that the test input string is rich enough to contain the state transition characteristics of all other input sequences. This is a reasonable assumption for this relatively simple problem.

ON THE CD

We have already reviewed system parity at the start of this chapter. When originally presented, the table for the parity machine was claimed to be equivalent to the finite state diagram. Now we can also see that this table is equivalent to Table 6.5 and thus should have the same behavior as the designed circuit. Pattern parity-verify on the CD-ROM contains the same input string as used in the verification example. Advancing the system shows that the behavior is also identical to that given in Table 6.6. After processing the entire string, the parity machine ends up in the even state, as desired.

A Sequence Recognition Example

In order to further illustrate the sequential design process, we reprise the earlier example of sequence recognition shown in Figure 6.2. Recall that the purpose of this finite state machine was to generate an output whenever the sequence "abac" was received. In addition, if a partial sequence that matched an initial sequence of letters in this string was received, rather than returning to the INIT condition, the state corresponding to this sequence would be entered.

As accomplished designers, we will not strictly follow the sequence of steps in the sequential design process, but rather keep within their spirit and at the same time attempt to expedite our efforts. Accordingly, we begin by redrawing Figure 6.2 as in Figure 6.7, with the appropriate binary designations substituted for both the states and the state transitions. There are five states; thus, three binary state variables are needed. The states are labeled with these variables in sequential, counting order. There are three possible transition symbols, "a," "b," and "c." It is possible to represent each with a unique bit but we can achieve some design economy by encoding the transitions with two bits only. Thus, we label the transition "a" with "00," "b" with "01," and "c" with "10."

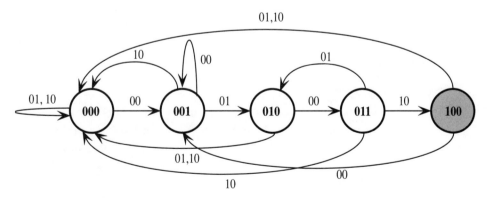

FIGURE 6.7 The DFSM in Figure 6.2 recast by creating binary designations for the states and state transition symbols.

At this point we can proceed directly to the excitation/output table in Table 6.7. This process combines state variable assignment, transition/output table construction, and excitation table construction. Again, the latter step is trivial when using D flip-flops and the flip-flop input variables, D0–D2 can be directly written on the bottom of the table.

TABLE 6.7 The Excitation/Output Table for the Sequence Recognition Example

	$I_0 I_1$			
$Q_0 Q_1 Q_2$	00	01	10	OUT
000	001	000	000	0
001	001	010	000	0
010	011	000	000	0
011	000	010	100	0
100	001	000	000	1
	$D_0 D_1 D_2$			

ON THE CD

System sequence2 on the CD-ROM is identical to sequence1 except that the former corresponds to the more compact variable encoding shown in Table 6.6. In this table, there are two input variables corresponding to the three input possibilities (the 11 case is not used). In the system, these variables are on separate rows of the input stream. Advancing through the system shows that it correctly recognizes the sequence 00 01 00 10, which corresponds to "abac." Loading pattern sequ2-pat2 shows that it also recognizes the sequence 00 01 00 01 00 10, corresponding to "ababac," as in sequence1.

To derive minimized excitation equations, Table 6.7 is transferred to the Karnaugh maps in Figure 6.8. Unlike the previous example, there are now three maps that must be minimized, one for each of the flip-flop inputs. As usual, care must be taken when placing entries on these maps. The table rows and columns are in counting order and the Karnaugh map conforms to the Gray ordering. There are two additional aspects of these maps that we have not previously encountered.

1. Representing five (or more) variables on a Karnaugh map

Each flip-flop input is a function of five variables: three state and two input variables. Previously, the largest Karnaugh map we have seen contained four variables. To represent more than four variables, we need to create multiple maps and split one or more variables by its possible values. For example, in this case, we assign $Q_0 = 0$ to the first map and $Q_0 = 1$ to the second (we could have chosen any other variable to split with equivalent results). If we needed to represent six variables, we would split two variables to create a total of four maps. It is possible to continue this process for an arbitrary number of variables although for greater than six variables, the minimization process would be very unwieldy. In these cases, algebraic methods are used instead, which are usually implemented in software in order to assist the digital designer.

On a five-variable map, if a prime implicant is repeated on both maps, then the two implicants can be joined. In describing the new prime implicant, the splitting variable is dropped. This is easily seen to be the same procedure we have always used for forming prime implicants and naming them (see Exercise 6.6). For example, in the map in Figure 6.8(B), the two prime implicants representing the four center cells are joined, and the description does not include the splitting variable Q_0. However, if the prime implicant was only on the right map, then the description would be $I_1 Q_0 Q_2$ instead of $I_1 Q_2$.

2. Don't care conditions

The x's on the maps represent unknown values or don't cares. In this example, there are two origins of these conditions. First, we do not know anything about what happens to states where $Q_0 \; Q_1 Q_2$ are 101, 110, or 111. Recall that this is a five-state machine represented by three state variables. The three unknown states arise because the three variables represent a total of $2^3 = 8$ possible states, three more than we need. In addition, we know nothing about the transitions between states when $I_0 I_1 = 11$. In these cases also, we enter x's in the table.

Because the x's represent nonexistent states or inputs, it is permissible to set any of these cells to either 0 or 1 (but not both), at our convenience, without changing the behavior of the machine. One proviso must be kept in mind, however. There is a small risk that this will lead to erratic behavior on the part of the machine if we happen to accidentally enter a forbidden state or use a nonexistent input (this may happen, for example, because of noise in the inputs or the system). It is safer to simply set all the x's to 0s. This way, if a nonexistent state happens to be entered, or if an aberrant input is present, the transition to the all 0 state will be affected. If this state happens to be the INIT state, as in this example, we are guaranteed that the machine will be effectively reset in these cases.

160 Digital Design: From Gates to Intelligent Machines

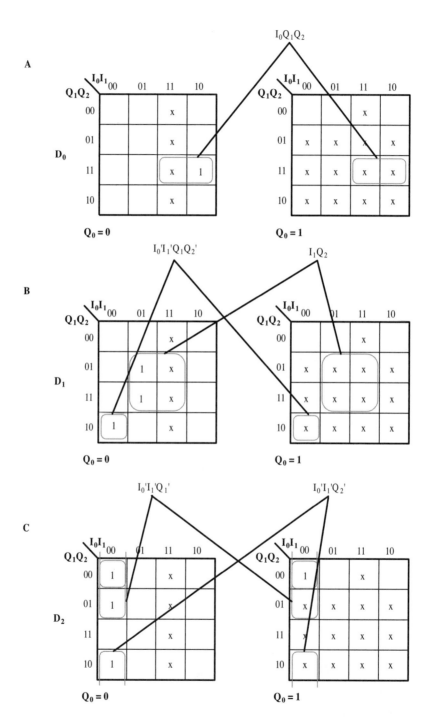

FIGURE 6.8 Karnaugh maps corresponding to each of the state variables in Table 6.7, with x's in the cells representing nonexistent inputs and unused states.

However, if we are willing to take the (usually small) risk that the machine does not behave in an unexpected fashion, we can exploit the don't cares in order to achieve greater minimization. This is illustrated in Figure 6.8. For D_0, there is only a single minterm present. Using the fact that we can set x's to 1s where desired, we can, however, create a prime implicant that covers more cells. This is accomplished by assuming that the x next to the prime implicant is 1, and that the two matching x's on the right map are also 1. Recall from the discussion of five-variable maps that this allows us to draw a single implicant that spans both maps, and that in the description of this implicant the splitting variable (Q_0 in this case) drops out. Thus, the final description of the prime implicant is $I_0 Q_1 Q_2$, as opposed to $I_0 I_1 Q_0 Q_1 Q_2$ if the don't cares were not used.

The prime implicants for the other maps are similarly formed, and the final expressions for each flip-flop input are given in Equations 6.6a through 6.6c.

$$D_0 = I_0 Q_1 Q_2 \tag{6.6a}$$

$$D_1 = I_1 Q_2 + I_0' I_1' Q_1 Q_2' \tag{6.6b}$$

$$D_2 = I_0' I_1' Q_1' + I_0' I_1' Q_2' \tag{6.6c}$$

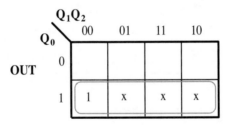

FIGURE 6.9 The Karnaugh map for the output with don't care states indicated with x's.

Using the same minimization procedure on the output (see Figure 6.9), the output can be reduced to a single variable:

$$OUT = Q_0 \tag{6.6d}$$

The circuit resulting from implementing Equations 6.6a through 6.6d appears in Figure 6.10. As usual, the next state logic is implemented with NAND gates, except for the first flip-flop, which requires a single product term only. To verify the correctness of the device, we will apply the string "bababacabacc" to the device. Referring to the original DFSM in Figure 6.2, we see that this machine should generate two outputs, one after the initial sequence "bababac" and another after the second appearance of the substring "abac." Table 6.8, which shows all inputs and state transitions and the translation of the binary inputs and states back to their original symbolic form for reference, verifies that this is indeed the case. As in all verification computations, the states Q at time step n are functions of the states Q at time step $n-1$ and the current inputs (we also assume that the system starts in the state $Q_0 = Q_1 = Q_2 = 0$).

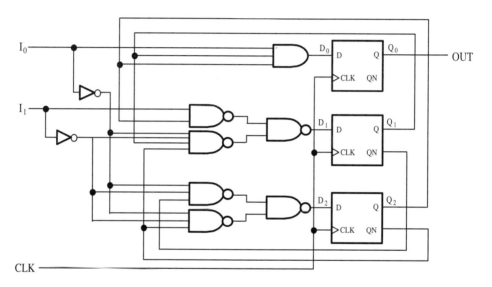

FIGURE 6.10 The realization of the sequence recognizer.

A Maze Example

The main point of interest in the following example, as in many real-world cases, lies not so much in the realization of the circuit given the DFSM (which, though complicated, is at its core a mechanical process) but in representing the problem in such a way that it can be subjected to the current formalism. The task is to build the logic that will instruct a robot to turn in the right directions in order to solve a maze without making a wrong turn. This must be realized by a sequential circuit, rather than a combinational circuit, because under identical circumstances the robot may need to perform different turns. That is, it needs to remember what has happened to it so far in order to make the correct turn at each junction.

Sequential Machines

TABLE 6.8 The Verification Table for the Sequence Recognizer

0	0	0	0	0	0	0	1	0	0	0	1	1
I_1	1	0	1	0	1	0	0	0	1	0	0	0
input	b	a	b	a	b	a	c	a	b	a	c	c
Q_0	0	0	0	0	0	0	1	0	0	0	1	0
Q_1	0	0	1	1	1	1	0	0	1	1	0	0
Q_2	0	1	0	1	0	1	0	1	0	1	0	0
state	INIT	a	ab	aba	ab	aba	abac	a	ab	aba	abac	INIT
OUT	0	0	0	0	0	0	1	0	0	0	1	0

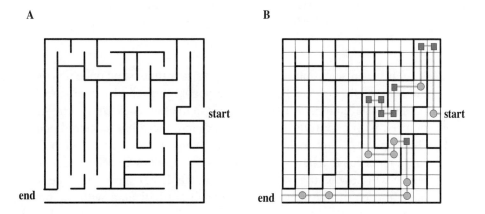

FIGURE 6.11 (A) The maze to be solved, and (B) the solution with the junctions marked. Choice points are represented by circles and junctions where only one turn is permitted are shown as squares.

Figure 6.11(B) shows the solution to the maze overlaid onto a grid with these junctions marked. The first step in representing the maze as a finite state machine is to make the distinction between the junctions for which there is a choice to be made and those in which only one turn is possible. These are marked by circles and squares respectively, in the figure. The various types of junctions are cataloged in Figure 6.12, which shows what the maze looks like *from the point of view of the robot*, as it travels in the direction indicated by the arrow. There are four no choice conditions. For example, in the first (labeled "000"), the robot sees a wall to the left and to the right and therefore must go straight (assuming a complete reversal of direction is not permitted), and in the second (labeled "001"), a wall appears in front and to the left and therefore the robot has no choice but to turn right. There are also three situations in which the robot must choose which way to go. For example, in the first such situation (labeled "100"), the robot can choose to travel straight or turn right.

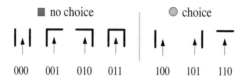

FIGURE 6.12 The various types of junctions cataloged as no choice and choice points. Each is assigned a unique binary code.

This representation can be used to form a finite state once it is understood that the state of the robot tells it how to negotiate each choice point (the nonchoice points, by definition, will always result in the same behavior). The simplest way to proceed is to go through the maze and note the robot's desired behavior at each choice point. If this behavior needs to change for a given choice point, it is necessary to proceed to a new state (otherwise, the robot will behave as it did previously, which will now be the wrong direction). Table 6.9 shows how the robot must behave at each of the choice points, which are indicated with the binary labels as given in Figure 6.12. The robot can stay in the same state up until choice point 6, because its behavior until then is consistent: right turns on "001" junctions, and left turns on "110" junctions. However, at point 6, it must go straight on "100." Therefore, we must change state at this point. Recall that behavior is a function of both the current state and current input, and if these are constant then the behavior will be constant. By allowing a state change, we are providing a mechanism by which the output behavior can change for identical inputs. As the table indicates, once this state change is affected, it is sufficient for the robot to stay in this new state until the end of the maze. Its behavior is consistent throughout this period.

In order to derive the finite state machine corresponding to Table 6.9, we note the following. The transition from state A to state B must be triggered by some aspect of the input, and the input right before the change occurs is "100." However, if we let the input alone trigger the change, it would be made on choice 1, not choice 5. Therefore, we must only change state when the input is "100," *and* after the robot has been moving left, at choice points 3 and 4.

TABLE 6.9 The Required Behavior of the Robot at Each Choice Point and the State Associated with Each Point

Choice	Type	Behavior	State
1	100	right	A
2	100	right	A
3	110	left	A
4	110	left	A
5	100	right	A
6	100	straight	B
7	110	right	B
8	100	straight	B
9	100	straight	B

This can be achieved with the addition of an extra state, as shown in Figure 6.13. The robot stays in the initial state, here labeled "00," on all initial turns. These include, for completeness, those at no choice points and choice points (we assume that the dead-end junction "011" is never hit and ignore this case). Note that unlike our previous examples, all transitions in the diagram contain an output behavior in addition to an input label. This is a consequence of the fact outputs are functions of inputs as well states. In other words, this is a Mealy rather than a Moore machine.

The additional new state is entered when the robot must make a left on "110." If we simply entered state B after the first left on 110 (at choice 4), then the next left on choice 5 would be wrong. The extra state thus allows the progression to what is labeled state B in Table 6.9 at the correct time, and not before.

In summary, an analysis of the turns necessary to solve the maze indicates that two distinct states are needed. However, in order to provide an unambiguous clue as to when the transition occurs, a new state needs to be created. The creation of extra states of this nature is a common feature of complex sequential circuit synthesis, and arises whenever a set of inputs and the current state alone do not uniquely determine the next state of the machine.

At this point, we can proceed to the comparatively easy task of translating Figure 6.13 into a design. We begin by translating the finite state machine into the excitation table shown in Table 6.10. Because this is a Mealy machine, there is not a separate column for the outputs. Rather, the outputs ("s" for straight, "l" for left, and "r" for right) appear in each cell, reflecting the fact that they are determined by the input sequence as well as the state.

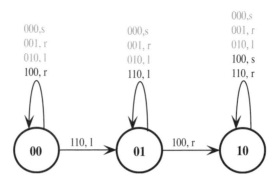

FIGURE 6.13 The DFSM for the maze problem. Nonchoice points are in gray and choice points in black.

TABLE 6.10 The Excitation/Output Table for the Sequence Recognition Example

Q_0Q_1	$I_0I_1I_2$				
	000	001	010	100	110
00	00,s	00,r	00,1	00,r	01,1
01	01,s	01,r	01,1	10,r	01,1
10	10,s	10,r	10,1	10,s	10,r
	D_0D_1				

The excitation/output table is converted into Equations 6.7a through 6.7d with the Karnaugh maps in Figures 6.14 and 6.15. The standard minimization procedure, over five variables (i.e., a double map), is used here with the following provisos. First, we can reduce the size of the final design by assuming that unless directed otherwise, the robot will move straight. Thus, there is no need to provide logic for this contingency. Second, additional reduction in design complexity can be achieved by assuming that the don't cares (the x's on the maps) can be set to 1s at our discretion.

$$D_0 = I_0I_1'Q_1 + Q_0 \tag{6.7a}$$

$$D_1 = I_0'Q_1 + I_0I_1Q_0' \tag{6.7b}$$

$$\text{Left} = I_1Q_0' + I_0'I_1 + I_0I_1Q_1 \tag{6.7c}$$

$$\text{Right} = I_2 + I_0I_1'Q_0' + I_0I_1Q_0 \tag{6.7d}$$

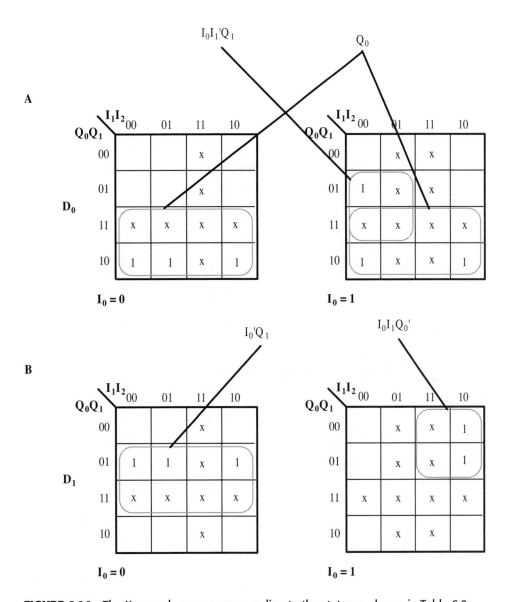

FIGURE 6.14 The Karnaugh maps corresponding to the states as shown in Table 6.8.

These excitation and output equations are rendered into the circuit shown in Figure 6.16 in the standard manner. The next-state logic is realized with a set of NAND gates, and the output logic is likewise implemented. One change with respect to earlier examples is that the output logic is quite a bit more complex. This

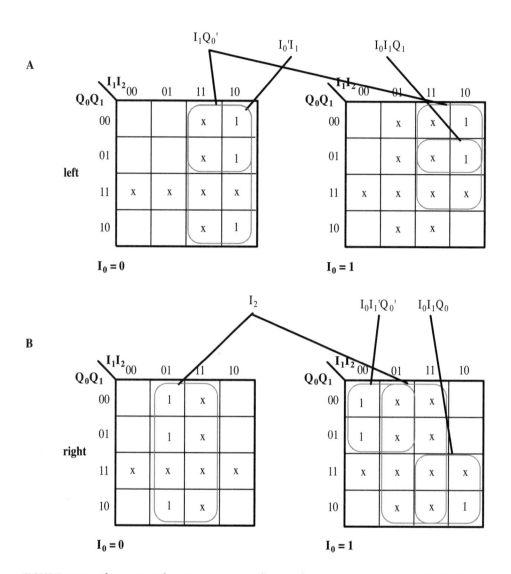

FIGURE 6.15 The Karnaugh maps corresponding to the outputs as shown in Table 6.9.

reflects the fact that Equations 6.7c and 6.7d include inputs as well as prior states. An additional minor change, albeit larger conceptual alteration, is replacement of the clock signal with the product of this signal and a cell detector. It makes little sense to have the robot's behavior triggered by a regular clock. Instead, we want the robot to consider changing direction only as it enters into a new cell, where each cell is an element in the grid shown in Figure 6.11(B). Only when this occurs should it consider changing its state.

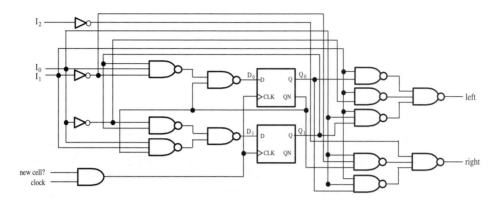

FIGURE 6.16 The circuit corresponding to excitation and output Equations 6.7a through 6.7d.

After all this work, it is natural to entertain doubts that the robot will work; thus, verification is more important in this example than in previous ones. Table 6.11 shows that the robot does indeed successfully negotiate the maze. As you work through the table, remember that inputs are given from the point of view of the moving robot and not as you view the maze from a static position. Also keep in mind that if neither left nor right is indicated, the robot moves straight by default.

TABLE 6.11 The Excitation/Output Table for the Sequence Recognition Example

turn point	0	1	2	3	4	5	6	7	8	9	10	11	12	13	14	15	16	17
input ($I_0 I_1 I_2$)	100	010	010	100	010	001	001	001	010	010	110	110	100	001	100	110	100	100
state ($Q_0 Q_1$)	00	00	00	00	00	00	00	00	00	00	01	01	10	10	10	10	10	10
input (as a picture)	↑	↑	↑	↑	↑	↑	↑	↑	↑	↑	↑	↑	↑	↑	↑	↑	↑	↑
output	right	left	left	right	left	right	right	right	left	left	left	left	right	right	str	right	str	str

LATTICE simulates this system (found in maze) on the CD-ROM by having separate machines for each of the outputs. The first component of this system, the state machine, reproduces the state transitions necessary to make the correct turn decisions. The other components are these decisions, one cell for the left turn, and one cell for the right turn. The system is constructed from the excitation/output table in Table 6.11. In accord with this table, there are three states, and the transition from 00 to 01 (turquoise to light blue) occurs when 110 is received as the input. The transition from 01 to 10 (light blue to blue) occurs when 100 is received. The left cell is designed to be active whenever the input is 010, or when the input is 110 and the state is not 10. The right cell is designed to be active whenever the input is 001, or the input is 110 and the state is 10. Work through each of the inputs in Table 6.9. As you can see, the system makes the correct turn decision in each case.

One of the many lessons regarding intelligent systems that have emerged over recent years is that complex behavior does not necessarily depend on a complex underlying mechanism. This example (the robot only has a 2-bit memory) as well as numerous others in this book exemplify this principle. In fact, one of the driving forces behind the interest in cellular automata is the idea that exceedingly simple rules generate exceedingly complex (and often beautiful) animations.

What enables this to happen is that much of the complexity lies not in the individual machine but in the environment. In the case of cellular automata, this complexity is embedded in the states of the surrounding automata, effectively augmenting the internal workings of the target automata with external memory and mechanisms. This effect is also seen in natural systems. For example, ethologists, who study animal behavior, have long speculated that phenomena such as the flocking behavior of birds and the foraging behavior of insects can be explained by simple rules operating in a complex and dynamically shifting environment. Just how far this idea can be carried in explaining intelligent behavior is still unknown, and many researchers remain skeptical that the environment contains the right kind of complexity to achieve this goal. It is noteworthy, for example, that a machine that could learn to solve any maze would be much more complicated than the one we described here. We will return to this topic in the final chapter.

DESIGNING WITH J-K FLIP-FLOPS

As you may recall from Chapter 5, J-K flip-flops have more functionality than D flip-flops in the sense that the state of the flip-flop can be placed in the opposite state with a single set of inputs (J = K = 1, see Table 5.4). This can, on occasion, lead to a more efficient design. Generally speaking, today J-K flip-flops are shunned despite this advantage because they require more transistors to implement and the design process is more difficult. Nevertheless, we present an example with these circuits as a means of introducing the extra step in the design process that is required whenever the next state of Q, Q^+, is not equivalent to D for a given flip-flop.

This is the case with the J-K flip-flop, which instead has the characteristic equation

$$Q^+ \equiv QN\,J + QK'. \qquad (6.8)$$

Given this relation, in order to set the flip-flop, that is, to set Q, one must set both J and K to the appropriate values. Table 6.12 summarizes each of the possibilities. For example, suppose the Q is now 0, and we wish to maintain Q in this state ($Q^+ = 0$). This corresponds to the first row of the table. It is easily seen from Equation 6.8 that it suffices to set J to 0, and it doesn't matter what K is. Then $Q^+ \equiv QN \equiv 0 + 0 \equiv K' \equiv 0$. The other three rows of the table are derived in an analogous fashion. To get Q to go from 0 to 1, set J to 1. To get Q to go from 1 to 0, set K to 1. Finally, to maintain Q at 1, set K to 0.

TABLE 6.12 The Values of J and K for Each Transition of Q from Its Current to Next State

Q	Q+	J	K
0	0	0	x
0	1	1	x
1	0	x	1
1	1	x	0

As an example of J-K synthesis, we reprise the original example from Table 6.1 that began this chapter, that of a 2-input mod 4 counter. The transition table for this example is reproduced in Table 6.13. When both inputs are 1, the state jumps forward by two states, when a single input is 1, the state advances to the next state, and when both inputs are 0 the state remains constant.

TABLE 6.13 The Transition/Output Table for the Mod 4 Counter Example

Q_0Q_1	I_0I_1 00	01	10	11	OUT
00	00	01	01	11	1
01	01	11	11	10	0
11	11	10	10	00	0
10	10	00	00	01	0
	$Q_0^+Q_1^+$				

In order to implement this problem with J-K flip-flops, we need to convert Table 6.3 into Table 6.4 with the aid of Table 6.2. This is accomplished by noting each transition and substituting the appropriate values of J and K. For example, let's examine the shaded cell in Table 6.14. The transition for Q_0 is from 0 to 1, as indicated in Table 6.3. From Table 6.2 we can see that to effect this transition, we need to set J to 1; we don't care about the input for K and thus write x. Likewise, the transition for Q1 is from 1 to 1; thus, we set J and K to be x and 0 respectively. The other cells are filled in accordingly.

TABLE 6.14 The Excitation Table for the Mod 4 Counter Using J-K Flip-Flops

Q_0Q_1	I_0I_1 00	01	10	11
00	0x,0x	0x,1x	0x,1x	1x,1x
01	0x,0x	1x,x0	1x,xo	1x,x1
11	x0,x0	x0,x1	x0,x1	x1,x1
10	x0,ox	x1,0x	x1,0x	x1,1x
	J0 K0, J1 K1			

Only after completing this extra step (which was unnecessary with D flip-flops), can we minimize for each of the J and K inputs to the flip-flops. Figure 6.17 shows the corresponding Karnaugh maps. The equations below are derived from these maps in the normal manner (see Exercise 6.9).

$$J_0 = I_0I_1 + I_1Q_1 + I_0Q_1, \tag{6.9a}$$

$$K_0 = I_0I_1 + I_0Q_1' + I_1Q_1' \tag{6.9b}$$

$$J_1 = I_0I_1 + + I_1Q_0' + I_0Q_0' \tag{6.9c}$$

$$K_1 = I_0I_1 + I_1Q_0 + I_0Q_0 \tag{6.9d}$$

J_0

Q_0Q_1 \ I_0I_1	00	01	11	10
00			1	
01		1	1	1
11	X	X	X	X
10	X	X	X	X

K_0

Q_0Q_1 \ I_0I_1	00	01	11	10
00	X	X	X	X
01	X	X	X	X
11			1	
10		1	1	1

J_1

Q_0Q_1 \ I_0I_1	00	01	11	10
00		1	1	1
01	X	X	X	X
11	X	X	X	X
10			1	

K_1

Q_0Q_1 \ I_0I_1	00	01	11	10
00	X	X	X	X
01			1	
11		1	1	1
10	X	X	X	X

FIGURE 6.17 The Karnaugh maps corresponding to Table 6.12.

Once the excitation equations are derived, implementation of the corresponding circuit is straightforward and analogous to the method used with D flip-flops. Each of the Equations 6.9a through 6.9d are SOPs and therefore the next-state logic can be constructed with NAND gates as in Figure 6.18. Note that each flip-flop has both a J and K input in addition to the clock. The total number of NANDs is 13, which is comparable to the 12 NAND gates in the D flip-flop implementation in Figure 6.4. In addition, that circuit required that the rightmost two NANDs have relatively high fan-in, and also required two extra inverters, so the J-K implementation may be preferred under certain circumstances.

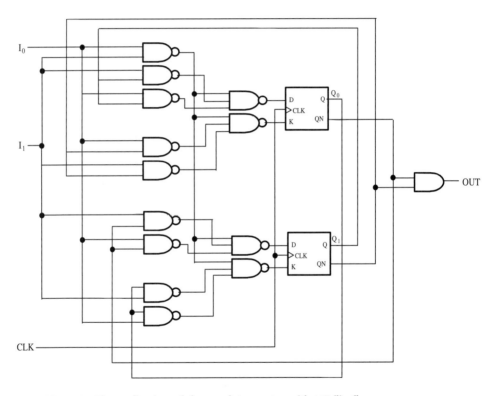

FIGURE 6.18 The realization of the mod 4 counter with J-K flip-flops.

SUMMARY

Unlike combinational circuits, the response of a sequential circuit is a function of its state as well as its current inputs. In order to model this type of process, DFSMs (deterministic finite state machines) were introduced. These models consist of a number of states with a transition leading to another state for each input combination. Finite state models are an excellent way to visualize the sequential process, but another and more convenient way to represent the same process from the point of view of sequential design is the state/output table.

Sequential circuit analysis consists, in essence, in the transformation of the circuit's behavior into such a table. In order to accomplish this, one must first derive the equations governing both the next-state logic and the output logic. From this, a transition/output table is constructed, and by labeling these states, the state/output table is produced. One may then optionally turn this table into a DFSM to help describe what the machine is doing.

State machine synthesis is a harder task. Perhaps the most difficult aspect of synthesis is the conversion of a word problem into a DFSM. In general, this aspect is nonmechanical and allows for a variety of solutions, some of which may be more elegant than others. The mechanical steps in synthesis run in the opposite direction of those for analysis. First, a state/output table is constructed. Binary designations are then assigned to each state. From this, an excitation/output table is constructed, which is trivial in the case of D flip-flops, but somewhat more difficult with other devices as the last example in this chapter demonstrated. From this table, minimized excitation and output equations can be derived which will form the basis of the finished design. It is especially important to verify the design once it is constructed in the case of sequential machines because errors may creep in at any stage of this lengthy process.

EXERCISES

6.1 Draw a finite state machine for each of the following cases:
  (a) A machine with a single binary input that generates an output when the sum of the bits received is an integer multiple of 3.
  (b) A machine to recognize the sequence ABBAAB (remember that any substring may form the start of the next string).
  (c) A machine to recognize the sequence xyzxxzzz.
  (d) A machine that generates an output only if doublets have been received in the suffix with the input characters "a," "b," and "c." Thus, on receipt of the last character, the strings "accbb" and "aaaabbcccc" would generate an output, but "abc," "aabbc," or "aabbccc" would not.

6.2 In the finite state diagram in Figure 6.2, what would be the consequence of routing all strings not on the path to the final string "abac" back to INIT. Give examples.

6.3 Reformulate Equation 6.4 with XOR and reimplement the circuit in Figure 6.6 with your new equation.

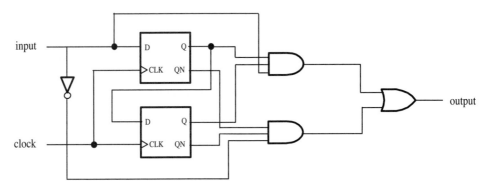

FIGURE 6.19 The sequential circuits for Exercise 6.4.

6.4 Analyze the following circuits and describe the recognition sequence in words for each:
(a) Figure 6.19
(b) A circuit with the following excitation and output equations (draw the circuit also):

$$D0 = I'Q0 + I\,Q1Q2$$
$$D1 = I'Q1 + Q1Q2' + IQ1'Q2$$
$$D2 = I'Q2 + I\,Q0'Q2'$$
$$out = Q0'Q1'Q2'$$

6.5 Create sequential machines corresponding to each of the DFSMs in Exercise 6.1.
6.6 Use standard simplification techniques to show why corresponding prime implicants on a five-variable map that are split along a variable into two four-variable maps may be considered a single product term.
6.7 Equation 6.7a states that D_0 is the sum of a product term and Q_0. Then why is the complement of this signal fed into the NAND gate leading to the first flip-flop in Figure 6.16?
6.8 Implement the parity function with J-K flip-flops. Compare the efficiency of the implementation with that in Figure 6.6.
6.9 Derive Equations 6.9a through 69d from the Karnaugh maps in Figure 6.17.
6.10 Design a sequential Mealy machine along the lines of the soda machine in the soda system. Include two outputs, one for the soda and one for the nickel return. In each case, the outputs should be a function of the inputs as well as the state of the machine.

LATTICE Exercises

6.11 Design a finite state machine with a single input that counts mod 3. Implement it in LATTICE.

6.12 Design a finite state machine that generates an output whenever an input string is of the form "caba*," that is, all strings containing "cab" followed by an arbitrary number of a's. Implement this in LATTICE.

6.13 Design a finite state machine that recognizes the username "abcb" and then the password "bbca." If the username is correct, it provides an output indicating this, and another output is shown when both user name and password are correct. Implement this in LATTICE. Why don't real password systems tell you if the username is correct?

ENDNOTES

1. This "trick" is not guaranteed to produce the minimal DFSM. There is a standard procedure to take any DFSM and reduce it to a minimal set of states, which is covered in most texts on automata theory.

7 Elements of Computer Design

In This Chapter

- Introduction
- Computer Organization
- Memory
- The CPU
- I/O
- Summary
- Exercises

INTRODUCTION

There can be no doubt that the computer is the jewel in the crown of the digital empire. It is worth taking a moment to understand why this is the case. Every other electronic device, and most mechanical ones for that matter, are designed for a specific task. A computer, on the other hand, is a general-purpose machine that can be configured to achieve an unlimited number of tasks. To take just one comparison, let us look at the difference between a relatively passive device such as a television and the computer. A TV is a complex electronic device—considerably more complex than anything we have looked at so far—but in essence, it takes an incoming signal, performs some minimal, and more importantly, fixed transformations, and displays the signal on a screen. In contrast, a computer can take a visual image and, depending on its programming, can perform any desired transformation. It can remove "red-eye" in a photograph. It can posterize the colors in the image (reduce the number of colors), or turn it into a black and white image at a lower resolution. It can reduce the image to a line drawing, or, with object recognition software of the kind we will consider in Chapter 9, pick out all the instances of men with bow ties and hats. In short, the capabilities of a computer are limited only by the ability of the programmer, not by the intrinsic wiring of the device.

Just how flexible are computers? The answer may surprise you. It is widely believed that even the most inexpensive of modern computers, given sufficient memory and time can perform any computation that can conceivably be realized by mechanical means. This hypothesis is difficult to formalize, and once formalized even more difficult to prove (and in fact, has not yet been proven in its most general form), but has a great deal of intuitive plausibility. Anyone who has programmed with a modern language, or even outdated ones, such as BASIC, quickly realizes that the language contains all the instructions needed to transform inputs into outputs in any desired fashion. Two aspects suffice to realize this: persistence on the part of the programmer, which is really a psychological barrier more than anything else, and as previously stated, an external memory. The latter may take any form: random access memory, a hard disk, or pencil and enough paper in the case of human simulations of algorithmic behavior. Those who wish to delve more into the intricacies of this idea can investigate the Church-Turing hypothesis and its variants, which are well described on numerous Internet sites.

Computers achieve this flexibility, of course, because they are not performing a fixed set of transformations from inputs to outputs but can be programmed to carry out an arbitrary transformation. Shortly we will consider the means by which this can be accomplished, but in the most general terms, the following must be in place for it to occur:

- There must be a means to iterate through a group of instructions. Instructions are typically stored in quickly accessible memory. These instructions tell the computer what to do at each time step. The normal procedure is for the computer to iterate through the instructions, executing one at a time, unless a branch is encountered (see the next condition).

- There must be a means of conditionally branching to another instruction. While it is conceivable that an algorithm for any finite task could be listed as a successive set of instructions, in most cases it would be extremely inefficient to write a program in this manner. What is needed is a means to branch to another instruction or set of instructions depending on a conditional test.
- The instructions must be complete relative to the task. It would do no good to have the proper flow of control through the instruction set if those instructions were not sufficient to perform the operations that are needed on the data. At the minimum, there is the need for elementary logical operations, such as AND, OR, and NOT (or just NAND) alone, and the ability to read and write from memory.

These are minimal requirements and most high-level programming languages (and to a lesser extent, machine languages) provide considerably more. For example, most languages have some version of a for loop, in which a block of instructions is executed a fixed number of times. This could be provided with a counter and a conditional test, but it is convenient to have this structure as a separate construct. Likewise, it is nice to have some form of subroutine that can be called repeatedly from multiple places in the program. These are flow-of-control constructs. It is also desirable to have instructions that carry out high-level operations. For example, although in principle it is possible to add two numbers logically, a computer language will rarely require a programmer to perform the addition this way. Normally an additional unit known as the ALU (Arithmetic Logic Unit) will perform this operation directly (see the next section). Normally, higher-level languages also contain a number of other arithmetic operations as well as an assortment of manipulations such as string transformations, bitwise manipulations of numeric quantities, and input/output instructions, including both file manipulation methods and graphical methods for displaying information on the screen.

We can summarize this discussion succinctly as follows. Unlike most digital circuits, sequential or otherwise, computers have the ability to interpret a set of instructions and to perform the actions indicated by these instructions. This separation between software (the list of instructions) and hardware (the computer itself) provides the flexibility to realize virtually any algorithm. Moreover, it is almost always much easier to change an instruction or set of instructions than to alter a hardwired circuit to realize a given task. In this chapter, we will lay the groundwork for the construction of such a system, and in Chapter 8 we will use this knowledge and that of previous chapters to construct a simple, but operational computer.

COMPUTER ORGANIZATION

Figure 7.1 shows the typical high-level organization of a computational system. It is comprised of three interacting components: the central processing unit, or CPU, which is the heart of the system, a memory subsystem that stores data for use by the CPU, and a number of peripheral I/O units for communicating the results of computations to the user and receiving inputs from the user.

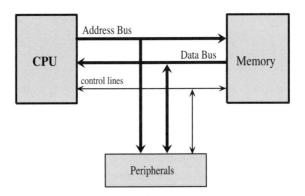

FIGURE 7.1 Three components of a computational system.

Before considering each component in detail, let us briefly consider how communication between the components takes place. The CPU determines what memories to access and which peripherals to address through the address bus. Recall from Chapter 5 that we can think of a bus as a collection of individual data lines. The size of this address bus will determine how many distinct memories can be stored. For example, if the bus were 2 bytes wide (16 bits), then the number of distinct memories that could be addressed would be 2^{16}, or one for each of the possible combinations of bit values on the bus. Some of these bits memories are also needed to specify the peripheral uniquely, if there are any being addressed at any given time. The small number of such devices relative to the size of the memory usually means that most of the addressing is devoted to nonperipheral memory operations, however.

The address bus is unidirectional, that is, it is sent only from the CPU to the memory subsystem and the peripherals. In contrast, the data bus is bidirectional. It transmits data from the CPU and to the CPU. For example, the CPU could be communicating with a writable CD player, in which case it needs to send and receive data from this device. The size of the data bus determines how much data can be transferred in parallel between the CPU and the memory or the peripherals. A 4-byte data bus means that up to 32 bits can be transferred at once.

The control lines comprise a number of individual lines with varied purposes. For example, they determine whether the CPU is accessing memory or an I/O peripheral, whether this access is a read or write, and whether an I/O device is ready to transmit data. Control lines can also be used to transmit interrupt signals, that is, signals telling the CPU to stop what it is doing and get ready to accept input. For example, rather than continuously looking at the mouse for input, which would be inefficient, most computers wait for the mouse to send an interrupt signal to the CPU telling it that there is new data to be processed. Although interrupts of this sort greatly increase the usability of the machine, we will ignore them for the purpose of simplicity in the following treatment.

Instead, we will concentrate on the essential operational aspects of computers in the following sections, starting with the memory subsystem, moving on to the basics of CPU design (the full design is outlined in Chapter 8), and then concluding with a brief treatment of I/O devices.

MEMORY

There are usually a variety of memory devices attached to a modern computer system. Arranged in the order of speed, these are registers (discussed in Chapter 5), a high-speed cache, random-access memory (RAM), the hard disk, and one or more high-capacity backup devices such as CD-ROMs, DVDs, or tape-based storage. The driving factor behind this hierarchy is cost.[1] While it would be ideal to a have large-capacity high-speed cache, this would be prohibitively expensive. Thus, typically there is a relatively small capacity memory of this sort located close to the processor (to minimize transfer speeds), and a much larger RAM. Similarly, it would be ideal to access only RAM rather than a slower hard disk. However, RAM at a comparable size of most hard disks would put the personal computer out of the cost reach of most consumers.

Here we will concentrate only on RAM, and leave the treatment of other forms of memory to textbooks dedicated to computer hardware. RAM is in a sense, the most essential form of memory, as we will see in Chapter 8, because it stores the software as well as the data that the software operates on.

Before pursuing this topic, it is necessary to introduce an additional device, the tristate buffer. We glossed over this type of circuit in earlier chapters, because its behavior is extra-logical, but it is now necessary in the context of reading memories. The tristate buffer is illustrated in Figure 7.2. This device consists of an input line, a buffer (the triangle), and an output line. However, unlike a normal buffer, say on an inverter, this buffer can be enabled or disabled. When it is enabled, the input signal passes on to the output line unhindered. When it is disabled, no signal is allowed to pass.

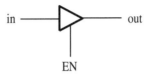

FIGURE 7.2 The tristate buffer.

Table 7.1 summarizes this behavior. The key entries appear when EN is inactive. In these cases, the buffer presents a high impedance, or high Z, to the output. What this means is that the tristate buffer acts as an open switch when it is disabled. A sufficiently high impedance or high resistance is functionally equivalent to an open switch, which offers infinite resistance. The purpose of such a device is twofold. First, as the table illustrates, it allows the input to propagate to the output only when enabled. Second, and just as crucially, it allows the outputs to be connected together to form the feed for a single line. This is illustrated in the case of reading from RAM, as discussed below.

TABLE 7.1 The Truth Table for the Tristate Buffer

EN	in	out
0	0	high Z
0	1	high Z
1	0	0
1	1	1

There are three primary requirements for RAM. First, it must be readable. Next, it must be writable, and unlike the PROM, writable many times. Third, and most crucial for the purposes of computers, both writing and reading must take place within the context of a supplied address. Only the memory contents at this address will be read, or in the case of writing, altered during the operation. Figure 7.3 shows the typical organization of a RAM device, in this case, one that stores four 2-bit memories.

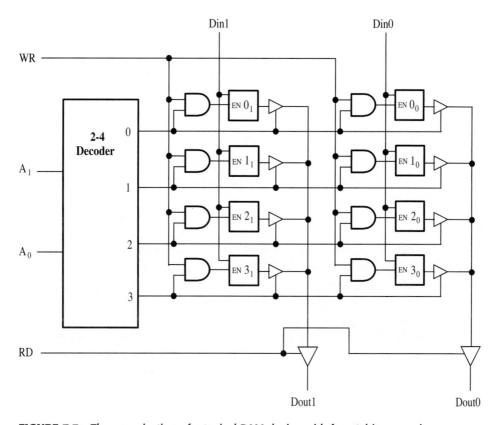

FIGURE 7.3 The organization of a typical RAM device with four 2-bit memories.

This is a relatively complex circuit, so we will consider the two operations (reading and writing) separately. Each corresponds to the activation of either the RD or the WR control signal, respectively. First, reading. For the purpose of illustration, let us assume that the second memory is being addressed, that is, $A_1 = 1$ and $A_0 = 0$. The n line two leading out of the decoder will be active and all other lines inactive. The only tristate buffers that will be enabled will be those corresponding to memory units 2_1 and 2_0. These will therefore place their values on the vertical lines leading to Dout1 and Dout0. If the read control signal RD is also active, then the tristate buffers for these lines will also be enabled, and the memory number 2 will be retrieved.

The purpose of the tristate buffers can now be made clear. From a logical point of view, an AND gate whereby the output of the decoder acts as a control line and the output of the memory unit acts as the data line would suffice. But then there would be no guarantee that current would flow from the memory unit to the output. Current on an active line would in this case, flow back into the AND gates serving the other outputs, and the current on the data line would be accordingly depleted (see Exercise 7.2). The situation with the tristate buffer is radically different, however. Recall that this buffer acts like an open switch when disabled. Thus, only the memory currently being addressed is connected to the output line at any given time, and this conflict is avoided. Likewise, we use tristate buffers with the RD control line as an enable because other signals may be placed on these lines in a full computational system.

Writing follows the same addressing procedure as reading. However, in this case, the output of the decoder is ANDed with the WR control signal to form an enable input to each memory unit. Once enabled, the unit is designed to be set to the signal that is present on the input lines. For example, suppose WR is active, RD is inactive, $A_1 = 1$, $A_0 = 0$, Din1 is 1, and Din0 is 0. Then memory unit 2_1 will be set to 1 and memory unit 2_0 will be set to 0.

Thus far, we have said little about the internal workings of the memory units. We will not explore this topic in detail because the physics of these devices is beyond the scope of this text and is not entirely germane to their *functional* properties. Nevertheless, we mention one key distinction, that between dynamic RAM, or DRAM, and static RAM, or SRAM. The former works like a leaky capacitor and must therefore be refreshed periodically. The latter works in a manner more similar to a register, and does not need refreshing. It is, however, more expensive than DRAM, and is usually reserved for the cache memory.

ON THE CD

System RAM in the Chapter7 folder on the CD-ROM contains an emulation of a 4 by 2 random-access memory. On the left is a 2-to-4 decoder; the current address is 10, which activates line 2. The write enable is active (top right), and we will first illustrate this activity. We are currently writing the bit string 10 (the contents of "in") to address 2. Stepping twice through the emulation first decodes the address ("addr2" becomes red). At the same time, the write enable to the left of each memory becomes red. Each memory is constructed to copy from the "in" machine when there is a red cell both to the left and the right. (Right-click on any memory unit to see how this is accomplished in the state table: variables I0 and I1 are defined to be 1 whenever there is a trigger of the color red anywhere to the left and right, respectively, of the memory unit.) Thus, in this case, the memory at address 2 gets written to.

Now enable reading by setting the read cell to blue. After another system advance, the memory that was just stored should be written to the out machine. This works because this machine first looks to the left to make sure that the read cell is set, and then looks to the right to determine which memory unit to read from. A similar system with a larger number of memories and more bits per memory will be used in Chapter 8 in the full computing system.

There are various means of increasing the capacity of a memory system that are independent of the type of memory. Suppose we require an 8 by 2 memory system instead of the 4 by 2 shown in Figure 7.3. One method of achieving this is illustrated in Figure 7.4. (for the purposes of illustration, only the read logic is shown). While it is perfectly possible to increase the size of the decoder to 3 to 8, and fill in the extra memory cells (see Exercise 7.4), this figure shows an alternative method. The 1 to 2 decoder decodes the lower-order bit. When the lower-order bit is 0, then the even memories, 0, 2, 4, or 6 will be addressed. When this bit is 1, the odd memories 1, 3, 5, or 7 will be addressed. The 2 to 4 decoder addresses the high-order bits, that is, it selects among these four possibilities. As an example, let us assume that the address 101 is applied to the memory. The high-order bits are 10, so the top decoder activates line 2. This selects either memory 4 or memory 5. However, the lower-order bits only allow memory 5 to be passed onto the outputs.

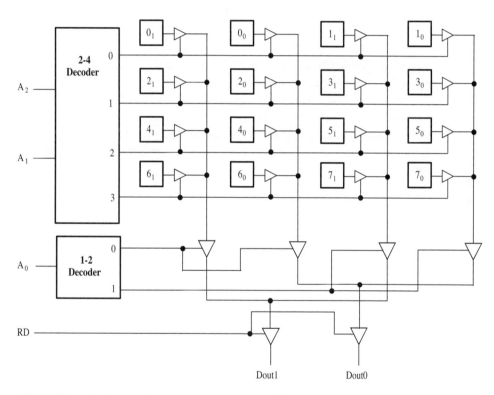

FIGURE 7.4 Implementing an 8 by 2 memory with two decoders.

Why go to the trouble of using two decoders when one will do? The answer has to do with the number of gates needed by the single decoder. You may recall from Chapter 4 that a decoder needs a single AND (or NAND) gate for each output. Let us say that we were using a 16-bit address, which is in fact small by contemporary standards. This would entail 2^{16} or 64 K gates for the decoder alone, an unmanageable number. The alternative is to use a two-dimensional structure as in Figure 7.4. One 8 to 256 decoder would address the high-order byte, and one 8 to 256 decoder would address the lower-order byte. The total number of AND gates needed in this case would then be 256 + 256 = 512. In other words, the original implementation requires approximately 128 times the number of gates (ignoring the extra tristate buffers that would be minimal relative to the number of AND gates in the single decoder case). This ratio becomes larger as the address grows, making it even more imperative to avoid the single decoder method when the memory system is large (Exercise 7.4 asks you to graph the number of AND gates for both cases as a function of the address size).

In addition to increasing the memory size, we would also like to widen the memory unit, or increase the number of bits per memory. This can easily be accomplished by addressing multiple chips in parallel, as shown in Figure 7.5. Once again, for simplicity, we illustrate reading only. The three address lines are connected identically to two 8 by 2 memory chips, which are shown by their logic symbol. Both chips are also fed by the same RD line, and also an enable line EN (the usefulness of this line will be demonstrated in the next circuit). The outputs are also identical, except that by fiat, we will designate the top two the high-order output bits and the bottom two the low-order output bits. In this manner, it is possible to combine two n by m chips to form an n by $2m$ circuit, or x number of n by m chips to form an n by xm chip.

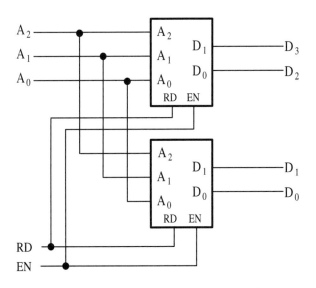

FIGURE 7.5 Creating an 8 by 4 memory with two 8 by 2 chips.

Finally, it is also possible to increase memory size by cascading memory chips in a manner similar to that in which we cascaded decoders in Chapter 4. The circuit in Figure 7.6 makes use of the enable line on two 8 by 2 chips to activate either the top or bottom chip to hold 16 memories. When line A_3 is inactive, the former is enabled and the output lines reflect the addressed content of this chip. When A_3 is active, only the bottom chip is enabled and the output lines reflect the content of this chip. Writing (not pictured) would be accomplished in a similar fashion to reading, with the write control signal entering in parallel to the two chips.

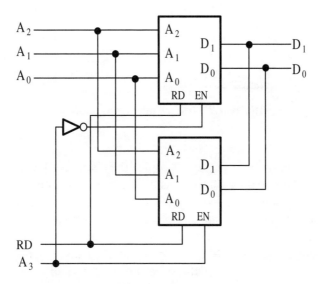

FIGURE 7.6 Creating a 16 by 2 memory with two 8 by 2 chips.

THE CPU

The design of a CPU will be treated more fully in Chapter 8. Here we will lay the groundwork for that chapter by introducing the overall organization of the CPU. Recall from the start of this chapter that a computer has three essential elements: the CPU, memory, and I/O. Likewise, the CPU itself comprises three elements, as illustrated in Figure 7.7: the control unit, the arithmetic logic unit (ALU), and a fixed set of registers.

If the CPU constitutes the brains of the computer, the control unit is the intelligence of the CPU. It is a finite state machine that determines what operations to execute and in what order. The control unit begins by fetching an instruction from

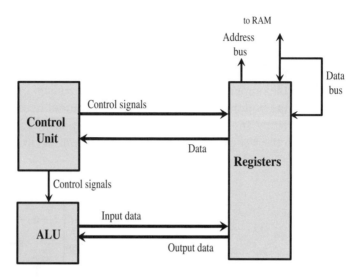

FIGURE 7.7 The overall organization of a CPU.

memory. In order to do this, it copies the contents of a register known as the program counter to the address register.[2] The program counter contains the location of the instruction to be executed. On the next cycle, the control unit causes this instruction to be read into a register known as the data register. The next step is to decode the instruction. This involves determining which of the many instructions the control unit should execute, as well as taking part in the instruction that contains the address of data to be operated on and transferring it to the address register.

What happens next depends on the instruction that was decoded. If it were an arithmetic or logic function, then the control unit instructs the ALU to implement this function. It accomplishes this by sending a set of control signals to the ALU, which causes it to perform the desired function. As an example, the ALU could take two bytes and return the bitwise ORed value of these bytes. The ALU receives its data from the registers and after its computation it sends the result back to the registers. It is also possible to have operations that do not involve the ALU. Among other things, these typically include unconditional and conditional jumps to addresses in memory containing the next instruction to execute. Jumps allow the program to depart from a strict sequential order of operation—a process, which as previously discussed, is essential to comprehensive software capability. Such instructions also include the ability to load items from memory (such as RAM) into registers, and to take the contents of a register and transfer them to a given location in memory.

In principle, registers could communicate with each other by special lines, but it is usually simpler to have the sending register place its contents on a shared data bus and have them read by other registers that have access to the same bus. This is indicated in Figure 7.7 by the recurrent connection from the register set to itself. The registers also send and receive data from memory, the location of which is determined by the contents of the address bus. Figure 7.7 also shows the other types of flow of information in general outline. As previously mentioned, the ALU reads from and writes to the register set. In addition, the control unit sends control signals to the ALU to indicate which operation it should perform, and to the registers, chiefly telling them when to be enabled. Finally, the control unit receives information from the instruction register and uses this information to determine what to do next.

If you are not confused by the previous discussion, you have not been fully paying attention! The only way to truly understand the operation of a CPU is to systematically work through a sample design, which is what we will do in the next chapter. For now, you should formulate the following general picture in your mind. The control unit fetches an instruction from memory, places it in a register, and decodes it, meaning that it extracts the instruction (and possibly some data that the instruction operates on) from the contents of that register. Then it executes this instruction, which means that it may tell the ALU to perform a given operation or may perform a direct operation on a register such as loading it into RAM. When this cycle is complete, it moves on to the next instruction, and this continues until the program terminates.

I/O

While it is possible to have separate busses and separate logic for I/O operations, it is increasingly common to share existing hardware when performing I/O. The reason for this is simplicity. Simplicity usually leads to greater design economy, which translates into a less expensive circuit, or alternatively, a circuit with the same expense but more capability. In this case, the economy arises because it is not necessary to have separate instructions for writing and reading to and from an I/O device. To write to the device, one simply performs a store, where the address specified is the fixed address of the device. To read from the device, one performs a load, again using the address of the device.

This is known as memory-mapped I/O because the external device acts as if it sits in a location in memory. There is some cost associated with this method. First, a dedicated section of memory must be devoted to this function. This is usually a minimal price to pay, given that RAM is now inexpensive and plentiful. The other price is that shared busses could result in conflicts. This cost can be minimized however, with proper design.

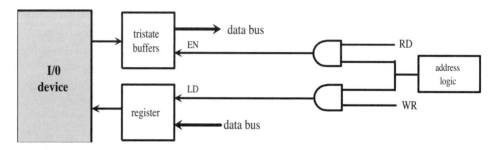

FIGURE 7.8 An I/O device and its interface with the existing computer architecture.

Figure 7.8 shows the interface between a typical memory-mapped I/O device capable of both reading and writing data, such as a hard disk, and the existing computer architecture. Reading involves first enabling tristate buffers. The enable is the product of a RD signal, generated by the CPU and logic that becomes active when the address that corresponds to the device is present. If we did not have this latter condition, the device would continuously write to the data bus whenever there was an RD control signal, whether this signal was intended for this device, another device, or a non-I/O-related read. When these two conditions are met, the output of the I/O device is placed on the data bus and then transferred to the relevant register (whichever register is indicated by the load operation). Likewise, writing to a memory-mapped I/O device is analogous to storing the contents of a register into a memory location. In this case, these contents are usually transferred to another register, buffering the result for further use by the device.

Figure 7.9 provides more detail regarding how this is accomplished in the case of a computer system with two I/O devices, one that is written to, and one that is read from. The former is associated with the four-bit address 1110 and the latter with the address 1111. The load line for the register is enabled when the first address is present on the address bus, and the WR signal is issued. This same address takes the memory at this location and places it on the data bus and this is loaded into the register. The first I/O device is now free to access this register as it needs to perform its given function. When the address 1111 is present, and the RD signal is active, then the tristate buffers on the second device are enabled. This places the current contents of this device on the data bus, and these are then transferred to the appropriate register.

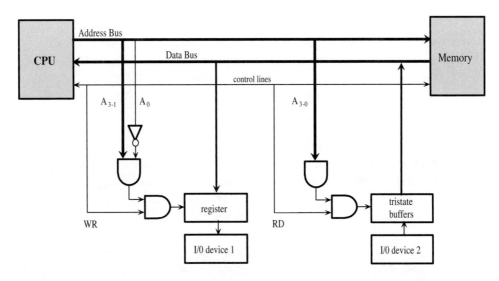

FIGURE 7.9 Two I/O devices, one that can be written to, and one that is read from, in the context of the larger computational system.

ON THE CD

System IO on the CD-ROM contains an emulation of an IO device that can be written to. This device takes the contents of a register (machine mem3 in the system) and encodes it into one of four notes, C through the next highest F. In order to do this, it must be enabled; the conditions for which are first that WRITE is active (this cell is red), and the second that the proper address for this device, 11, arrives on the address bus (EN is red). In the default configuration, the device is not enabled because the address is set to 01. Left-click on A1 in the address string, and then advance the system. This will enable the output device, which will then allow the contents of the register to reflect the data in memory mem3, which in turn will produce a note. Changing the contents of this memory while the device is still enabled will produce other notes. We will use a similar output device in the next chapter in order to illustrate how a computer can influence the external world.

SUMMARY

A computing system can be conceptually divided into three interacting units: the CPU, a fast memory store (usually RAM), and one or more I/O devices. The CPU is the heart of the system and comprises three components: the control unit, the register set, and the ALU, each of which is described in detail in Chapter 8. The task of the CPU is to read an instruction from memory, to decode that instruction, and to execute it. Depending on the instruction, this may entail telling the ALU (through control signals) how to operate on register contents, loading or storing from memory, or the flow of control operations such as conditional or unconditional jumps. It is by virtue of the execution of a variable, external set of instructions such as these that the computer system derives its unique flexibility relative to other digital circuits.

Memory is constructed from register-like cells that can be written to and read from. To write to a given location in memory, the address for that location must be decoded and this signal ANDed with the write control signal to enable this operation for a particular cell. Reading also consists of decoding the given address, but in this case the output of the decoder enables a tristate buffer so that only one cell at a time is connected to an output line. A two-dimensional memory structure is usually more efficient than a single dimensional one in that the total number of gates in the two decoders for the former is a fraction of the number of gates in the single decoder necessary for the latter. Memory chips may be cascaded to provide wider memories by feeding the enable of this chip with the high-order bit.

I/O devices are crucial for interacting with the user. The most common type of interaction with I/O was discussed here, the so-called memory-mapped I/O. In this case, a device is treated by the CPU just like a location in memory. To read from a device, the CPU simply reads from this address and to write to it, it writes to this address. The decoded address is also used as a control signal to enable the desired device. Together with the other components, the CPU, and the memory, a working computational system can be constructed.

EXERCISES

7.1 Figure 7.10 shows three additional tristate buffers. Construct the truth table for each given the following descriptions:
   (a) Figure 7.10(A): An inverting tristate buffer
   (b) Figure 7.10(B): A tristate buffer that is active-low enabled
   (c) Figure 7.11(C): An active-low inverting tristate buffer

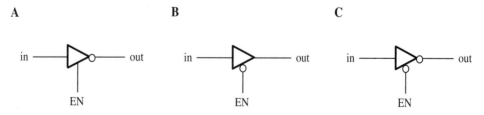

FIGURE 7.10 Three additional tristate buffers.

7.2 Replace the tristate buffers connected to the leftmost column of memory cells in Figure 7.3 with AND gates by using the output of the decoder as a control line and the output of the cell as the data line. What is the problem with your new design?

7.3 Implement an 8 by 2 memory system with a single decoder (show only the read logic).

7.4 Graph the number of gates needed for 1- and 2-dimensional designs for memories with 1 K, 4 K, 16 K, 64 K, and 256 K items. What can you conclude from your graph?

7.5 Construct a 16 by 1 memory with two 2 to 4 decoders.

7.6 Construct an 8 by 8 memory with four 8 by 2 chips.

7.7 Construct a 32 by 2 memory with two 16 by 2 chips.

LATTICE Exercises

7.8 Design a rudimentary control unit that works as follows. It consists of a single cell machine and two adjacent cells containing the current instruction. It begins in a state called "FETCH," then it proceeds to a state called "DECODE" regardless of the contents of the cells. Then it proceeds to one of four states, "I_0" through "I_3," depending on the binary value in the instruction cells. Each one of these states then returns to "FETCH," and the cycle repeats (this is a foreshadowing of the control unit we will design in the next chapter).

7.9 Realize Figure 7.4 in LATTICE by modifying the RAM system.

7.10 Modify system I/O by adding an additional I/O device mapped into address 01. This could also be a subsystem that plays music (for example, drum beats or short loops).

ENDNOTES

1. There is also the issue of volatility, or the ability to store memory when the power is off. The main memory of a computer is erased when the machine is turned off; the hard disk, of course, is not. However, volatility is less important than speed. For example, if RAM were not so much more expensive than disk space, computers would be constructed to always supply power to this form of memory so that it would be maintained.

2. In this and future discussions, we will assume that we are using the LATTICE computer, or L-puter, a vastly simplified computer introduced in Chapter 8. There is tremendous variation in the basic operations of a CPU, as well as other factors touched on briefly at the end of Chapter 8 that may cause, for example, the registers to be different or the operations to encompass more functionality. Nevertheless, the treatment here remains faithful to the core of the operation of every machine from laptop to supercomputer.

8 The Design of a Simple CPU and Computer

In This Chapter

- Introduction
- The Register Set
- The Instruction Set
- The Control Unit
- The ALU
- Putting it All Together
- Further Issues in Computer Design
- Summary
- Exercises

INTRODUCTION

This chapter introduces the most ambitious digital circuit in this book, that of a comprehensive, albeit simplified, computer. We will term our working model the L-puter, short for LATTICE computer. As discussed in Chapter 7, the CPU is the heart and soul of a computer, and is comprised of three components: the register set, the control unit, and the ALU. Each of these is now treated in turn. In addition, we will introduce the notion of an instruction set, which contains the list of operations that are understood by the CPU. Upon completion of the CPU design, we will then add memory and an I/O device to form a full-fledged working machine, on which we will run three test programs, each constructed from the collections of instructions.

THE REGISTER SET

Before describing the register set used in the L-puter, we will examine the structure of memory in this machine. This structure will influence both the size of the registers, and the overall design of the CPU as a whole. Each word in memory will consist of 7 bits. If the memory contains data, such as an argument for an addition, then it will be undifferentiated. If the memory contains an instruction however, it will be broken down as in Figure 8.1. The first three bits contains the instruction code, each combination of which uniquely encodes an instruction. Thus, the L-puter contains $2^3 = 8$ instructions, as described in the next section. The last 4 bits optionally contain an address (not every instruction makes use of these). This address contains the location of another item in memory, such as the location of a piece of data to load into a register. The L-puter is capable of addressing $2^4 = 16$ unique locations—tiny by contemporary standards, but sufficient for our purposes.

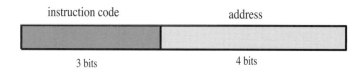

FIGURE 8.1 The structure of memory in the case of an instruction.

We can now better understand the structure of the registers in the L-puter as detailed in Table 8.1. There are a total of six registers in the design, four which are internal and for use by the CPU only, and two of which are programmer accessible. The first of the former type is the address register. This is designed to hold a 4-bit string that encodes a memory address. As we will see later, this address will be stripped from the instruction as a whole (all 7 bits in Figure 8.1) and placed in this register. The next register used by the CPU is the program counter. This holds a pointer to the instruction currently being used by the processor, and is likewise 4-bits wide, to address an instruction in any of 16 places in memory. The data register is 7-bits wide and therefore can hold the contents of an entire memory location. This register will be operated on to produce the address used by an instruction, if any, and to extract the instruction proper. This last 3-bit item will be stored in the instruction register, and the control unit will determine its course of action based on the contents of this register.

TABLE 8.1 The Register Set for the L-Puter

Register	Symbol	Description	Programmer Accessible
address register	AR	4-bit register storing memory address	no
program counter	PC	4-bit register storing instruction address	no
data register	DR	7-bit register storing instructions and data	no
instruction register	IR	3-bit register storing an instruction	no
accumulator	AC	7-bit special register for storing data	yes
general register	R	7-bit general register for storing data	yes

The final two registers are for use by the instructions, that is, their contents are affected directly by the program that is running. The accumulator is a special register that is read in every ALU operation, as described in the section below on this topic. The accumulator will also be used to store the results of all ALU operations, and in addition, as the basis for a conditional jump. The general purpose register R will contain the second argument in a two-argument arithmetic or logical operation. These two registers are the bare bones minimum required for the construction of a CPU. Most processors have many more, in order to increase the speed of operation. Writing to and from registers to RAM is a relatively time-consuming process, and therefore, the more data that can be held in registers, the swifter the processing time. Of course, there is a price to pay for more registers including increased complexity of design and a concomitant increase in the size and expense of the processor. As always, the onus is on the designer to determine the right trade-off between increased speed and the extra expense.

With a single exception, the registers can be implemented as in Chapter 4, that is, as a collection of flip-flops that can be loaded in parallel. However, there is an extra requirement for the PC register. This register, which holds the address of the current instruction, needs to be incremented by 1 after each instruction is processed so that the CPU can progress to the next instruction. While it would be possible to do this using the adder in the ALU, this would effectively entail that each instruction take as much time as two instructions: one time slice to implement the instruction itself and one to increment PC. Instead, what is usually done is to implement special-purpose incrementing hardware for this register.

One method of realizing such a counter is inspired by the following observation. When incrementing a binary number, the rightmost bit always alternates. For all others, the bit becomes 1 whenever the exclusive OR of that bit and the product of the bits to the right in the prior number (the number that is being incremented), is 1. For example, let us say we have counted to 0111. Then the leftmost bit goes to 1 because (0 XOR (1 · 1 · 1)) = 1; the third bit goes to 0 because (1 XOR (1 · 1)) = 0; the second bit goes to 0 because (1 XOR 1) = 0; and the rightmost bit goes to 0 because it flips. Thus the next number in the sequence is 1000. You can easily verify that the next numbers generated by this rule are 1001, 1010, etc.

Figure 8.2 illustrates a 4-bit counter using these rules and based on the D flip-flop. The output of each flip-flop feeds an XOR gate, which is then fed back into the flip-flop. This represents the first argument to the XOR. The other input to the XOR is the product of the outputs of the flip-flops above, representing the second argument. In the case of the first XOR, the input is simply logic 1 (Exercise 8.1 asks you to explain why this is the case). On each successive clock uptick, the outputs of the flip-flops represent the next 4-bit binary number. It is also possible to AND the clock with an enable so that the counting only proceeds when desired. Finally, each flip-flop also includes the ability to be set directly, by including a line that is ORed with the output of the XOR gate. This feature is especially useful when the flip-flops form a register, as is the case with the program counter PC, because the register needs to be set in addition to being incremented.

ON THE CD

System counter on the CD-ROM implements a sequential counter with an enable in a similar manner to the circuit in Figure 8.2. Each advance of the simulation adds 1 to the count, until all 4 bits are active and the count starts over. Clicking on a count cell Cn reveals how these machines work. The least significant bit reverses value on each advance. The others are the XOR of the previous value in the same cell and the product of the cells to the right, when the enable is active (in the other cases they simply maintain their previous state). Because this is a state table

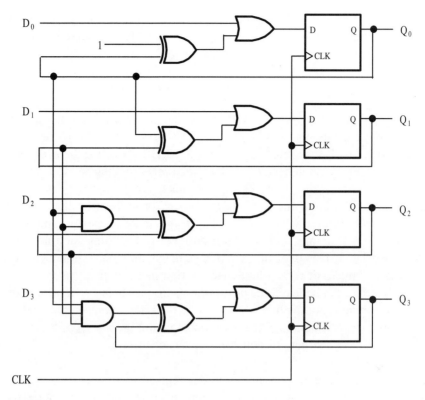

FIGURE 8.2 A 4-bit counter constructed from four D flip-flops.

rather than a truth function, the XOR function is represented implicitly by the cases where the function should be 1, rather than an explicit formula. Exercise 8.9 asks you to justify the values for each cell given the formula.

THE INSTRUCTION SET

As has been stressed repeatedly, the crucial difference between a computer and other digital circuits is that the computer can be programmed to follow a list of instructions. These instructions are usually in the form of a high-level language such as Java or C. However, the instructions in these languages are too complex to be interpreted by the processor per se. Take a typical line of code from a hypothetical accounting program written in Java:

```
totalPrice = totalPrice + price;
```

This augments the variable totalPrice by the amount contained in price. In order for this operation to work, the values of these variables must be retrieved from RAM, they must be sent to the ALU to perform the addition, and the result must be rewritten back to the location corresponding to the totalPrice.

Each of these subtasks corresponds to an elementary processor operation, and thus it would be awkward for the processor to interpret the high-level instruction directly. Rather, what typically happens is that processors come with a set of primitive instructions, into which high-level instructions such as the one above are translated. This translation may take place all at once, in which case the language is compiled, or it may take place at runtime, in which case the language is interpreted.[1] In either case, the processor needs only to understand how to work with the primitive set of instructions known as the instruction set, that is, how to take these instructions and translate them into an unambiguous set of tasks.

These instructions come in five basic varieties:

1. Move the contents from one register to another.
2. Load a register with the contents of a memory location (from RAM).
3. Store the contents of a register into a memory location (to RAM).
4. Request that the ALU operate on one or more registers, and optionally the contents of memory.
5. Jump, conditionally or unconditionally, to an instruction in a given memory location.

Of these, only the first is dispensable in the sense that it can always be realized by first moving the contents of a register to a prespecified memory location (an operation of type (iii)), and then taking these contents and loading them into the target register (an operation of type (ii)). The others types are absolutely necessary. The contents of registers must be loaded from memory and stored back into memory if the computer is to act on the contents of memory and if these actions are to be recorded. It must be possible for the CPU to instruct the ALU to perform an operation on the contents of a register and possibly the contents of a memory location; otherwise, it would just be reading and writing the same values back and forth from RAM. Finally, as previously argued, there must be the option to conditionally jump to a new instruction in RAM. This allows the machine to depart from a strict sequential step through the program, and is essential for higher-level constructs such as iterative loops.

TABLE 8.2 The Instruction Set for the L-Puter

Instruction	Instruction Type	Instruction Code	Operation
LDA	ii	000AAAA	AC ← M[AAAA]
STA	iii	001AAAA	AC → M[AAAA]
LDR	ii	010AAAA	R ← M[AAAA]
JNZ	v	011AAAA	IF (AC ≠ 0) goto AAAA
ADD	iv	100XXXX	AC ← AC + R
SUB	iv	101XXXX	AC ← AC − R
IOR	iv	110XXXX	AC ← AC OR R
AND	iv	111XXXX	AC ← AC AND R

Table 8.2 shows the instruction set for the L-puter categorized by the four essential instruction types. As with all L-puter components, the instruction set is a considerably simplified relative to the modern PC but remains representative of the larger instruction set found in these machines. The instruction code for each instruction is also listed. This contains the binary code for the instruction as it is stored in memory. For example, a typical instruction might be 0101101. The first three bits of this string, 010 encode the instruction name, which in this case is LDR. The last four bits, 1101, encode the address for this instruction. In this case, the entire instruction means that the contents of memory at location 1101 are loaded into register R. This is also indicated in the last column of the table, which describes the operation. The conventions for this column are that the direction of the arrow indicates the direction of the data flow, and M[AAAA] means that the memory at address AAAA (the As will be either 0 or 1 in a real address) is being referred to.

For example, the first instruction LDA loads the contents of memory at location AAAA into the accumulator. This is represented symbolically on the right of the table by the contents of memory at this location, M[AAAA], being written into the AC register. Note that memory is being addressed by 4 bits. As previously mentioned, this means that we will be capable of addressing 2^4 or 16 unique memory locations in L-puter. This is considerably smaller than most PC addresses, which are at least 32 bits wide, and can therefore address $2^{32} = 4$ gigabytes or more, but it is sufficient for illustrative purposes. The second instruction is the mirror image of the first: it takes the contents of the accumulator and places them in the memory addressed by AAAA. The next instruction, LDR is identical to LDA except that the writing is done to the general-purpose register R instead of AC.

Loading to R is an absolute necessity given the way in which the last four instructions, ADD, SUB, IOR, and AND work. Notice that these instructions take two operands, that in AC, and that in R, they perform an operation and place the result back into AC. The four instructions include two arithmetic operations (binary addition and binary subtraction) and two logical operations (inclusive OR and AND). These instructions will be carried out in the ALU, which as its name implies, is designed for both arithmetic and logical tasks. Note that no memory operation occurs during any of these operations; hence, the memory section of each instruction is ignored.

The last instruction that requires explication is the most complex. JNZ, standing for "jump if not zero," inspects the contents of the accumulator AC. If this register is zero, it does nothing. This will effectively allow the CPU to move on to the next instruction, as we will see below. However, if it is not zero, JNZ will cause a jump to the address AAAA. In a real CPU, there will be more conditional instructions of this sort, although this single instruction will prove sufficient for our purposes.

In the next section we will further break down each instruction into the role it needs to play, not just with respect to the programmer-accessible registers, but also with respect to the internal registers of the machine. Eventually, when we have designed the entire machine, we will create small programs for our newly constructed computer from these eight instructions.

THE CONTROL UNIT

The control unit decides what the CPU is doing at all times. First it instructs the CPU to fetch the next instruction, then it tells the CPU to decode it, and then it directs the execution of this decoded instruction. This cycle—fetch, decode, and execute—is considered first in this section, and the register operations for each of these is delineated. Next, we consider how the control unit moves from state to state in this cycle. The ordinary path is from fetch to decode, but after that there are many branches, depending on the instruction decoded. Finally, we consider the purpose of these state transitions, which is to set the control signals on the registers, the memory, and the ALU in a manner consistent with the current state of the control unit. This is discussed in the context of the data path, which describes how these components communicate with each other.

The Fetch-Decode-Execute Cycle

Every CPU follows some variant on the fetch-decode-execute cycle. The details may vary from machine to machine, and in most modern computers these operations may overlap in order to gain speed (see the discussion of pipelining, below), but the general order of events is fixed. First the control unit must fetch the instruction held in the address stored in the address register AR. Then it must decode the fetched instruction, which means both determining the nature of the instruction and also separating it from any data associated with the instruction. Finally, it executes the instruction and then returns to fetch a new instruction. This instruction will be the subsequent instruction in memory if there is no jump, or the instruction at the desired address if there is a jump.

In the L-puter, fetching consists of two sets of operations, which we will term F1 and F2. F1 consists of copying the address in the program counter to the address register in preparation for the actual reading of the instruction. Formally, we represent F1 as follows:

$$F1: \quad AR \leftarrow PC$$

F2 performs the actual read, by copying the contents of the memory location given by AR to the data register DR. In addition, we will take this opportunity to update the program counter PC. We need to do this sometime before executing the next instruction and after reading the current contents of PC; this is as good a time as any, as this operation and the memory read may be done in parallel. If PC needs to be modified as a result of a jump, we will handle this later in the cycle. Together, these operations can be represented as follows:

$$F2: \quad DR \leftarrow M \quad PC \leftarrow PC + 1$$

This completes the fetch part of the cycle. Now we must decode the instruction currently residing in DR. As Figure 8.1 and Table 8.2 indicate, an instruction in the L-puter has two parts. The first three bits encode the instruction proper, and the last four optionally contain data for the instruction to operate on. We will place these two pieces in separate registers for separate treatment, as described in D, the decode part of the cycle:

$$D: IR \leftarrow DR[6..4]$$

$$AR \leftarrow DR[3..0]$$

The notation [x..y] means that bits x through y only are copied. Thus, the instruction register (IR) receives only the three high-order bits of DR. AR receives the last four bits in preparation for a read from the indicated address.

This completes the fetch and decode part of the cycle. At this point, each instruction must be examined individually for its effects on the register contents. We begin with the load from memory and store to memory instructions, LDA, STA, and LDR. Each is described in terms of register activity in E1 to E3 respectively:

$$E1: AC \leftarrow M$$

$$E2: M \leftarrow AC$$

$$E3: R \leftarrow M$$

Each of these operations is predicated on the correct address, corresponding to the word in memory that is written to or read from, already having been placed in AR during the decode cycle. The astute reader should be wondering how the CPU distinguishes between these three operations, as they all involve transfer to or from memory. The precise answer will be given in more detail below. For now it is sufficient to know that the instruction that is decoding and now residing in IR will trigger different controls signals, read and write, and either AC enable or R enable, depending on these contents, and thereby effect only the correct transfer.

Then next instruction, JNZ, is a bit trickier in the sense that there are two possibilities. We first need to assume that the control unit has some way to detect that AC is not zero. This is not difficult to do, and a method for doing so will be described later. When this condition is met, we want to execute the following:

$$E4: PC \leftarrow AR$$

This replaces the previous updated PC (recall that it was incremented by 1 in F2) with the address to jump to. If the condition is not met, that is, AC = 0, then we simply take no operation. In that case, PC will reflect the contents of the instruction in memory following the current one, which is what we desire.

The last four instructions are relatively easy to implement. These are veridical copies of the instruction operations themselves, and are listed below:

$$E5: \quad AC \leftarrow AC + R$$

$$E6: \quad AC \leftarrow AC - R$$

$$E7: \quad AC \leftarrow AC \text{ OR } R$$

$$E8: \quad AC \leftarrow AC \text{ AND } R$$

The proper action in each case will be determined by control signals generated by the control unit and sent to the ALU. This is also described below.

In summary, there are two operations corresponding to the fetch cycle, one to the decode cycle, and then eight to the execute cycle. Only one of the eight, however, will be implemented for each instruction. The manner in which this is accomplished is described in the next section.

The Control Unit Finite State Machine

We can now reveal the punch line, so to speak, of all our hard work and preparation: the heart of the control unit is a device that we already have great familiarity with—the finite state machine. This unit needs to traverse from one state to the next of the fetch and then the decode parts of the cycle. Then, depending on the value of IR, it executes the particular instruction indicated by this register. There is only one other deviation, depending on the value of AC, and that is only for the JNZ instruction. All of these operations can be construed as the conditional movement from one state to the next; hence, the finite state formalism.

Figure 8.3 shows the DFSM for the control unit of the L-puter. Thick transition lines represent no choice transitions (or equivalently, transitions that take place for every input), and thin lines indicate transitions based on particular inputs. The initial state is the first fetch state, F1, and the unit always proceeds from here to the second fetch state, F2, and then to decode, D. The transitions from D depend on the contents of the instruction register IR. Each of the eight possible instructions corresponds to one combination of the 3-bit instruction code. The special case, corresponding to the jump instruction E4, involves a further test. If the sum of the bits in AC is 0, then there is nothing to execute; the CPU moves to the next instruction. If however, the sum is nonzero, then E4 is executed. In all cases, control returns to F1 after each instruction is executed, and the cycle repeats.

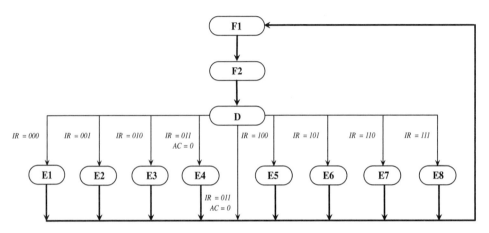

FIGURE 8.3 The finite state machine for the control unit.

ON THE CD

The control cell in system control on the CD-ROM represents the finite state machine in Figure 8.3. The table corresponding to this machine shows that the state transitions are a function of four variables, three corresponding to IR and one to the sum of the bits in AC. Despite the large number of cells in the table, it is remarkably simple. F1 always proceeds to F2, F2 to D, and then the instruction is selected by the contents of IR. For example, the default pattern in the system takes the control unit to E6 and then back to F1. The one exception is the case where the instruction is 011 and the sum bit is 0. Here the control unit returns directly to F1 without executing an instruction (the jump to another instruction is not performed). This can be seen by opening the pattern control-pat2, which takes the control unit to E4. Now render the sum unit inactive by right-clicking on it. The control unit skips E4 and returns directly to F1.

It is not difficult to convert the finite state machine in Figure 8.3 into an excitation table in preparation for realizing this machine. Table 8.3 shows the result (in

TABLE 8.3 The Excitation Table for the Control Unit

		Σ $IR_0IR_1IR_2$								
State	$Q_0Q_1Q_2Q_3$	0000	0001	0010	0011	1011	0100	0101	0110	0111
F1	0000	0001	0001	0001	0001	0001	0001	0001	0001	0001
F2	0001	0010	0010	0010	0010	0010	0010	0010	0010	0010
D	0010	0011	0100	0101	0000	0110	0111	1000	1001	1010
E1	0011									
E2	0100									
E3	0101					0000				
E4	0110									
E5	0111									
E6	1000									
E7	1001									
E8	1010									
		$D_0D_1D_2D_3$								

this table, the "Σ" is not a summation per se but a variable indicating that the sum of the bits in AC is not zero). In the table, states F1 through E8 have been converted to binary code. There are eleven states; thus, four binary variables and therefore four flip-flops are necessary. The remarkable simplicity of the table is a consequence of most states returning to F1 in an unqualified fashion. This is represented by the large "0000" at the bottom of the table. Otherwise, F1 always moves to F2, F2 to D, and D to the various instructions, or directly back to F1 as the case may be.

Without this simplicity, the control unit would be unwieldy in its realization. Each D is a function of eight variables, and this would be difficult to render on a Karnaugh map (although the minimization could be carried out with the appropriate software). However, given the relatively few transitions, Table 8.3 can be converted directly to the excitation Equations 8.1a through 8.1d, without the need for Karnaugh minimization:

$$D_0 = Q_0'Q_1'Q_2Q_3'\Sigma'IR_0IR_1 + Q_0'Q_1'Q_2Q_3'\Sigma'IR_0IR_1'IR_2 \tag{8.1a}$$

$$D_1 = Q_0'Q_1'Q_2Q_3'\Sigma'IR_0'IR_1'IR_2 + Q_0'Q_1'Q_2Q_3'\Sigma'IR_0'IR_1IR_2' + Q_0'Q_1'Q_2Q_3'\Sigma IR_0'IR_1IR_2 + Q_0'Q_1'Q_2Q_3'\Sigma'IR_0IR_1'IR_2' \tag{8.1b}$$

$$D_2 = Q_0'Q_1'Q_2'Q_3 + Q_0'Q_1'Q_2Q_3'\Sigma'IR_1'IR_2' + Q_0'Q_1'Q_2Q_3'\Sigma IR_0'IR_1IR_2 + Q_0'Q_1'Q_2Q_3'\Sigma'IR_0IR_1'IR_2 \tag{8.1c}$$

$$D_3 = Q_0'Q_1'Q_2'Q_3' + Q_0'Q_1'Q_2Q_3'\Sigma'IR_2' \tag{8.1d}$$

For example, D_0 is 1 in only three cells: when $Q_0Q_1Q_2Q_3$ is 0010 and the input variables are 0101, 0110, and 0111. We can combine these last two, and along with the Q conditions by standard simplification rules, this yields the first product term of Equation 8.1a; the second product term corresponds to the first cell. The derivation of the other equations is left as an exercise for the reader, as well as the construction of the circuit corresponding to these equations.

Data Paths

The control unit dictates what the CPU is doing at each stage. The remaining piece of the puzzle concerns what the control unit is supposed to be controlling. Fundamentally, this circuit generates control signals for the registers, for the memory, and for the ALU. When done correctly, this will effectively implement each of the states, which in the case of the L-puter, are F1, F2, D, and then one of E1 through E8.

Before describing the actions of the control unit in detail however, we need to provide a more detailed picture of the interaction between the elements that this unit affects. Figure 8.4 illustrates one possible architecture for the L-puter. Each of the registers reads from the data bus. A bus, as you may recall, is a collection of shared lines. Rather than having a direct connection between each of the interacting registers, the bus acts as a central highway from which the registers can draw data as needed. In this case, the bus is 7 bits wide, comprising the 3 instruction bits and 4 address bits in each instruction. The data is read into the register from the bus when the respective load signal is active. For example, to load DR, DRLOAD is set active, and the contents of the bus are then placed in this register.

Some of the registers are also capable of writing to the bus. In this case, there are multiple tristate buffers between the register and the bus, in order to prevent multiple outputs connecting to the bus at the same time. These buffers also contain an enable, which is set by the control unit when writing from the register to the bus is called for. For example, when register DR needs to place its contents on the bus, the control signal DRBUS is made active.

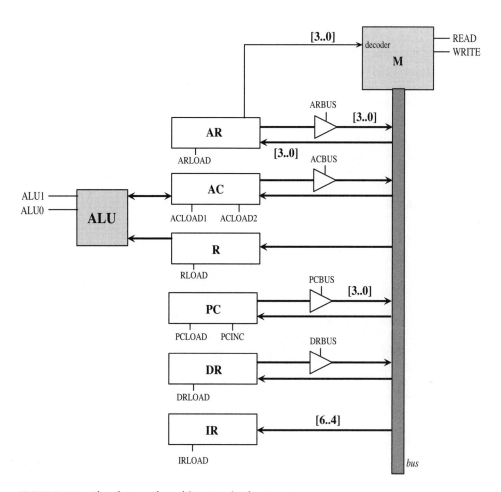

FIGURE 8.4 The data path architecture in the L-puter.

There are two remaining aspects of this architecture that require discussion. The first, the ALU, is treated in depth in the next section. The other aspect is the memory subsystem (M in the figure). Memory is always addressed by the four least significant bits in AR. That is, the contents of AR determine which of the $2^4 = 16$ memories is being addressed at any given time. When the READ signal on M is asserted then the contents of the given address are placed on the bus. When the WRITE signal is asserted, the contents of the bus are written to the indicated address.

We now systematically work through each of the states of the control unit in order to determine which control signals must be active at each stage; the results of this analysis are summarized in Table 8.4, which also repeats the register transfers previously derived. Fetch state F1 requires that the program counter PC be copied into AR. In general, a register to register transfer will first entail writing the contents of the register read from to the bus, and then transferring these contents to the destination register. In this case, we first assert PCBUS so the contents of PC are transferred to the bus, and also assert ARLOAD so that AR can read these contents.

TABLE 8.4 Active Control Signals for Each State of the Control Unit

Control State	Register Transfers	Control Signals
F1	AR ← PC	PCBUS ARLOAD
F2	DR ← M PC ← PC + 1	READ DRLOAD PCINC
D	IR ← DR AR ← DR	DRBUS IRLOAD ARLOAD
E1	AC ← M	READ ACLOAD2
E2	M ← AC	ACBUS WRITE
E3	R ← M	READ RLOAD
E4	PC ← AR	ARBUS PCLOAD
E5	AC ← AC + R	ACLOAD1
E6	AC ← AC − R	ALU0 ACLOAD1
E7	AC ← AC OR R	ALU1 ACLOAD1
E8	AC ← AC AND R	ALU0 ALU1 ACLOAD1

F2 requires that the addressed contents (that is, the contents of memory at the address now held by AR) of memory be read from M. Thus, we assert READ, which transfers the addressed memory to the bus, and then DRLOAD, which picks up these contents and transfers them to DR. In F2 we also assert PCINC, to increment contents of this register. D requires that DR be written to both IR and AR. Therefore, we assert DRBUS to write the contents of DR to the bus, and then IRLOAD and ARLOAD to read these contents. In the case of IR, we assume that it is hardwired to only pick up the three high-order bits; AR can load as normal and will simply ignore these high-order bits.

The rest of the entries in the table correspond to the executable instructions. E1 involves a transfer from memory to AC. Thus, we assert read and then ACLOAD2 (the other load for AC is ACLOAD1 and will be described shortly). The flow of data for E2 is the reverse of E1. Here we assert ACBUS to load AC onto the bus, and then WRITE to place these contents in memory (again, as addressed by AR). E3 is identical to E1 except that RLOAD is asserted to load to register R. E4, corresponding to the jump instruction, transfers AR to the bus, and then transfers these contents to PC. Extra hardware, described below, adds the facilitating condition to this transfer that AC is not zero.

In order to describe the ALU instructions, one perhaps counterintuitive concept needs to be stressed. No harm will be done if the ALU carries out its operations regardless of whether it is explicitly called to do so. The only time the ALU affects the contents of AC is if the control unit is in one of the states E5–E8. Thus, the ALU is continuously reading the contents of AC and R, but only writes back to AC during these states. As you can see from Table 8.4, each is therefore accompanied by ACLOAD1, which transfers the result of the ALU operation back into AC. The other control signals dictate which of the four ALU operations are chosen; this is described in more detail in the next section.

Each of the control signals can be formed as the sum of the control states that assert this signal. For example, READ = F2 + E1 + E3. That is, READ is asserted in control unit states F2 or E1 or E3. Likewise, ACLOAD1 = E5 + E6 + E7 + E8 (Exercise 8.7 asks you to supply the conditions for the other control signals). These conditions are of course trivial to implement and an OR gate will suffice for each signal.

ON THE CD

System data is by far the most complex system we have seen to this point, but mastery of this is essential to understanding the final system in this chapter, which describes the L-puter in full. On the left are the registers as shown in Figure 8.4, and on the right are 16 7-bit memories. Each register is accompanied by the relevant control bits. In addition, there is a bus that can be written to from the appropriate register, or memory, as the control bits to the left and right of this machine are set. Memory is addressed by the four low-order bits in the machine addr, which is simply a copy of the AR register. A special placeholder M holds the contents of the current memory being addressed at any given time. When READ is asserted, this is transferred to the bus. When WRITE is asserted, the contents of the bus are transferred to the memory that is currently addressed. This address currently selected for is indicated by a select bit (in orange) next to each memory.

The easiest way of getting a feel for the operation of this system is to work through a few operations corresponding to the states of the control unit. As Table 8.4 indicates, F1 involves a transfer of the contents of PC to AR. To do this we assert PCBUS (PCBS in the system) and ARLOAD (ARLD). This is already done in the default configuration of the system, where the former is set to yellow (state H) and the latter to lime (state J). Advancing the system will first copy the contents, 101, of PC to the bus, and then transfer it to AR. As a side-effect of setting AR, the contents of memory location 5 will be loaded into M (after three more cycles).

Now let's begin the next state, F2, in which the contents of memory are transferred to DR. As Table 8.4 indicates, we need to assert READ and DRLOAD. To carry out the former, set PCBUS (PCBS) to inactive (gray, state A), and READ (READ also) to saturated yellow (state H). For the latter, set DRLOAD to saturated yellow (it should already be in this state). After advancing the simulation twice, DR will hold the original contents of memory 5, 0110011, as desired. As you can see, setting the control signals to their proper values is an intricate task; however, this will normally be done by the control unit, as we will see in the final system in this chapter. Our task at that point will both less cumbersome and more entertaining—that of programming the L-puter by placing the appropriate instructions in memory.

It remains to connect the finite state machine to the data path architecture in order to realize a working CPU. However, before doing so, we introduce the final and crucial component of this system, the ALU.

THE ALU

If the control unit is the brains of the CPU, then the ALU is the brawn. Without an ALU, a CPU would be virtually impotent, unable to manipulate the data in RAM in any significant way. Here we will realize an ALU that is capable of carrying out the four arithmetic and logical instructions in the instruction set, which are repeated in Table 8.5 for clarity.

TABLE 8.5 The Four Arithmetic and Logical Instructions in the L-Puter Instruction Set and the Corresponding Control Signals

Instruction	Type	ALU 1	ALU 0
ADD	arithmetic	0	0
SUB	arithmetic	0	1
IOR	logical	1	0
AND	logical	1	1

The first two instructions are arithmetic and the last two are logical. The job of the ALU is twofold. First, it must be capable of carrying out each of the individual operations separately. Generally speaking, a unique circuit must be designed for each operation, although some economy may be gained in blending two operations. We will see an instance of this, for example, in the case of the arithmetic operations.

The next task is to send the desired result back to register AC. This result will be selected by the control signals sent by the control unit to ALU (in this case, there are two such signals encoding each of the four instructions). There are two conceptually distinct ways to carry this out. The first way is to perform only the instruction that was encoded on the control signals. The second method is to generate all results, but only place the desired one back to register AC. While both are functionally equivalent, the latter is simpler when all the hardware to carry out each instruction is already in place. We will not be saving anything by performing only the indicated instruction, and in fact the situation is slightly more complicated in the sense that only the correct subcircuit must be enabled. Therefore, we will generate all the results in parallel (with the exception of addition and subtraction, as explained below), and select the answer we want depending on the desired instruction.

Figure 8.5 shows how this can be accomplished in this situation. This circuit takes advantage of the fact that control signal ALU 1 selects between the arithmetic and logical operations, and ALU 0 selects among the two instructions in each subset. Suppose, for example, that ALU 1 and ALU 0 are 1 and 0 respectively. Then the mulitplexer (MUX) on the right receives input from both the logical sum and logical product of the AC and R registers, but only the sum is sent to the MUX on

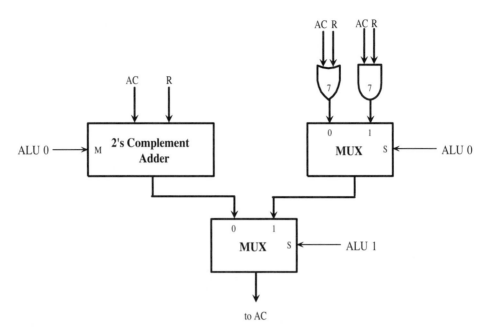

FIGURE 8.5 The design of the ALU for the L-puter. Both the AND and OR gate symbols stand for a collection of respective gates that process 7 bits in parallel.

the bottom. This MUX then selects for the result of the logical operations and sends this register back to AC. Likewise, if the control signals are 1 and 1, then the logical product will be routed to AC. Note that both the sum and product are computed in parallel; however, in the end, the accumulator receives the desired result only.

The situation is a bit more complicated with the arithmetic instructions. To understand what happens in this case, it will help if you refresh your memory by reviewing the operation of the two's complement adder and subtracter pictured in Figure 4.6. You may recall that this circuit had a dual role. When M was 0, it operated as an adder. However, when M was 1, all the bits in the second argument to the adder were flipped, and in addition, the initial carry was set to 1. This process turned this argument into a negative number, and in effect, turned the adder into a subtracter. By letting M take the value of ALU 0, we can therefore switch between an adder and subtracter as needed. In this case, there is no need for separate hardware for these two instructions.

The net result of the two MUXs and the control signal for the two's complement adder is to select for the desired result. This result can then be written back to the accumulator, and the operation is complete. Like all the other CPU elements in this chapter, this ALU is comprehensive, albeit simplified, relative to the ALU in a typical working computer. All ALUs, however, no matter how complex, contain a number of distinct operations that are selected by the control signals sent by the control unit. Other issues that would arise were we to design a full-fledged ALU, include multiplication, division, and the treatment of floating-point numbers in the case of arithmetic operations, and bit-shift and additional logical operators such as XOR in the case of logical operations.

System ALU on the CD-ROM implements a 7-bit arithmetic-logic unit along the lines of Figure 8.5. There are distinct machine sets for each of the OR, AND, and two's complement adder circuits, and as in the figure, each runs in parallel. It is the job of the ALU signals to select among the results, and in the case of the adder, to switch between addition and subtraction. In the default configuration, ALU1 is active (pink) and ALU 0 is inactive (yellow). The result machine looks to its left and observes that this is the case, and selects the result of the OR operation by copying this logical sum to itself. The OR operation works by copying the AC and R registers to OR1 and OR2, respectively, and then setting the result to active (blue) whenever at least one of these two bits are active. Now set ALU1 to red, or active (ALU1 and ALU0 have different active colors so that the result machine can distinguish between the two), and advance the simulation. Notice that the result of the AND operation is now placed in the result. AND works in a similar manner to OR, except of course it requires that both of its arguments are active before becoming active.

To test the adder, which works along the lines of the two's complement adder we constructed in Chapter 4, set both ALU1 and ALU0 to inactive, or yellow. This will set both M and the initial carry to inactive, and thus cause the adder to perform addition. The arithmetic sum is then copied into the result when the addition is complete after six iterations. The answer is 0110001, or 49, which is the correct sum of the contents of AC, 0101011, and R, 0000110. Now set ALU0 to active, or red. In this case, both M and a cell that is a copy of the ALU0, ALU0(c), become red. The former effect causes the contents of ADD(2s) to become the complement of ADD2, which is a copy of register R. The latter sets the initial carry bit to 1. Together, these effects turn the adder into a subtraction device, as can be observed by iterating through the system. The eventual result is 0100101, or 37, the correct result of subtracting R from AC as you can readily verify. This adder does not detect overflows (Exercise 8.10 asks you to provide such a detector).

PUTTING IT ALL TOGETHER

All the parts are now in place, and it is possible to describe how to create a full-fledged, working computer. Figure 8.6 illustrates the overall design. This diagram is essentially the register architecture from Figure 8.4 with three key additions: a clock, the control unit, and memory-mapped I/O. The control unit has already been described as an isolated circuit implementing a finite state machine, but here it is included in the architecture as a whole. The task of this unit is to decide on the next state according to the contents of IR, which after executing cycles F1, F2, and D, will contain the current instruction. Then, depending on the state, it will assert one or more control signals. For example, if IR contains the string 000, indicating E1, the control unit will assert READ and ACLOAD2 (refer to Table 8.4). It is by virtue of asserting the appropriate control signals at the appropriate time that the control unit directs the CPU into the desired course of action.

The clock serves two purposes. First, as in all digital sequential circuits, both the registers and memories load synchronously. Thus, they require a clock signal to direct their actions. Typically this will either be the uptick or downtick of this signal. The clock also plays another important, though less obvious role. Not every instruction takes the same amount of time. For example, loading the contents of memory into AC takes two clock cycles, one to load memory onto the bus, and one to write to AC per se. However, ALU operations may take as long as the data word is wide. In the worst case, for example, an adder without look-ahead adding two 8-bit numbers could take eight cycles, one for each of the carry bits to propagate.

For this reason it is necessary that the clock provides an additional indication to the control unit telling it when it is safe to proceed to the next instruction. The simplest, although not necessarily the most efficient method of doing so (see the discussion on pipelining below), is to set the wait time in each cycle to the same time it takes the longest possible instruction to complete. If, for example, 8 clock ticks were required for addition, the system would only jump to the next state after eight such time periods, regardless of how long a given instruction takes. This wait applies to all states, including fetch, decode, and all instruction states.

The final aspect of the complete design in Figure 8.6 is the addition of I/O at the bottom of memory. In other words, this machine contains memory-mapped input and output devices as discussed in Chapter 7. This means that to write to an I/O device, it is sufficient to write to a particular address in memory, and to read from that device, it is sufficient to read from that particular address.

To illustrate how the L-puter functions as whole, we will look at its operation when processing three programs. The first, and simplest, is given below:

```
LDA     13      ; load AC from address 13
STA     14      ; store AC to address 14
```

The Design of a Simple CPU and Computer 219

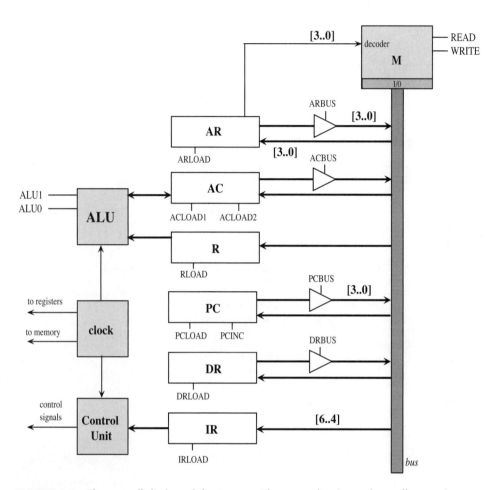

FIGURE 8.6 The overall design of the L-puter. The control unit conducts all operations by setting the appropriate control signals at the appropriate times.

This program takes the contents of memory location 13 and loads them into register AC, and then takes these contents and stores them in memory location 14. This program can be deconstructed as in Table 8.6, with the help of Table 8.4. The analysis assumes that the first instruction is located at address 0000 and is 0001101 (i.e., load AC from 1101) and the second instruction is located at address 0001 and is 0011110 (i.e., store AC into 1110), and the contents of address 1101 are 1010101.

The real work of the program is done by E1 and E2; the other stages F1, F2, and D are preparatory for these actions. Once the address for location 1101 is extracted from the first instruction and placed in AR, E1 can issue the command to load the contents addressed by AR into AC. Likewise, once the address for location 1110 is extracted from the second instruction, E2 need only issue the command to copy AC into this piece of memory.

TABLE 8.6 The Flow of Control, Register and Memory Transfers, and Register and Memory Contents after These Transfers, for the First Program

Control State	Register and Memory Transfers	Register and Memory Contents	Comments
F1	AR ← PC	M[0000] = 0001101 M[0001] = 0011110 M[1101] = 1010101 AR: 0000000	Initial continents of memory are shown; AR now has the address of the first instruction.
F2	DR ← M PC ← PC + 1	DR: M[0000] PC: 0000001	DR contains the first instruction; PC is incremented by 1.
D	IR ← DR AR ← DR	IR[6..4] = 000 AR[3..0] = 1101	DR is loaded into IR and AR. The three high-order bits of IR contain the instruction code for E1 and AR contains an address for this instruction.
E1	AC ← M	AC = 1010101	AC is loaded with the contents of M[1101].
F1	AR ← PC	AR: 0000001	AR is loaded with the incremented PC value and now points to the second instruction.
F2	DR ← M PC ← PC + 1	DR: M[0001] PC: 0000002	DR contains the second instruction; PC is incremented by 1.
D	IR ← DR AR ← DR	IR[6..4] = 001 AR[3..0] = 1110	DR is loaded into IR and AR. The three high-order bits of IR contain the instruction code for E2 and AR contains an address for this instruction.
E2	M ← AC	M[1110] = 1010101	The contents of AC are copied to M[1110].

ON THE CD

System Lputer on the CD-ROM contains an emulation of the L-puter. The system mirrors Figure 8.6 and is a composite of the earlier systems control, data, and ALU. The control unit, a single cell labeled CNTL is located in the lower left and its importance to the system is belied by its size. It is responsible for driving all of the control signals that surround it, which in turn drive the register loads and stores. The control unit itself is driven by the clock above; it is only permitted to change state every 10 clock cycles. This is to give the ALU, in the worst case, enough time to generate its result and send it back to the ALU.

The flow of control follows Table 8.6. The first instruction at location 0000 is loaded onto the DR and then placed in both IR and AR. The astute reader may realize that the actual writing to and retrieving from DR can be skipped and the instruction simply placed on the bus and read into IR and AR from there. However, in general, it is a good idea that a separate register holds the instruction. It may be used for further processing; in addition, there is the possibility that the bus will be used for other operations, such as interrupts (see the next section), and then the instruction will be lost when the CPU resumes its normal sequence of events. Although AR and IR receive the full instruction, not all of it is used. In the case of AR, it is first copied to the address machine, and then only the four least significant bits are copied to cells a3 to a0, which are used to decode the memory address for reading and writing. In the case of IR, only the three most significant bits are copied to cells IR6 to IR4, which are then used to direct the control unit to proceed to one of the states E1 to E8.

The first instruction in M0 is 0001101, the first three bits of which, 000, code for instruction E1 (load AC from memory), and the last four bits, 1101, contain the address that is read from. Stepping through the simulation, you can see the contents of this address being loaded into AC. The system then proceeds to the next instruction at 0001, the store instruction. This is decoded in the same way, and when the control unit reaches state E2, WRITE is asserted on the memory (maximize the cell window to see this). With address 14 selected, it writes the contents of the bus to this address, and the program is complete.

Our next program, which tests the operation of the ALU, is given below:

```
        LDA     13      ; load AC from address 13
        LDR     14      ; load R from address 14
        ADD             ; add AC and R
        LDR     12      ; load R from address 12
        SUB             ; subtract R from contents of AC
        LDR     11      ; load R from address 11
        IOR             ; OR R with contents of AC
        LDR     10      ; load R from address 10
        AND             ; AND R with contents of AC
        STA     14      ; store final result in address 14
```

Let us assume that the initial contents of addresses 10, 11, 12, 13, and 14 are 1010110, 0111000, 0000010, 1010101, and 0000001 respectively. The program first takes the contents of address 13 and loads them into AC, and then takes the contents of address 14 and loads them into register R. It adds these two registers, and as with all ALU operations, the result is written back to AC. The contents of AC should then be 1010101 + 0000001 = 1010110. The contents of address 12 are then loaded into R and then subtracted from the new contents of AC. Thus, AC should now hold 1010110 – 0000010 = 1010100. This is then ORed with 0111000, and AC should then hold 1111100. Finally, this result is ANDed with 1010110 to yield 1010100.

ON THE CD

Reading pattern lput-pat2 from the CD-ROM into the Lputer system allows one to run the current program. Stepping through the simulation reveals that AC receives correct values in succession: 1010101, 1010110, 1010100, 1111100, and 1010100. This last value is then written to memory address 14. Maximize the animation window to see the ALU at work. Note that this unit is always calculating something. However, it is only on an ALU instruction that the result is placed back into AC.

The L-puter is a didactic tool, but it is also a simulation tool. One of the reasons for performing a simulation before actually constructing an actual machine along these lines is to reveal possible faults. One problem immediately becomes apparent when running this program—there is no stopping condition. Left to its own devices, the system will begin to process data as instructions. Thus, on a real machine, we would need to add a mechanism to halt the program. This would be one of the many responsibilities of the operating system, which, among other things, interacts with the user to load and terminate programs.

Our final program is constructed to test two further aspects of the system. The first aspect is memory-mapped I/O. Let us map address 15 into an output device. As discussed in Chapter 7, this means that anytime a word is written to this address it will be sent to the I/O device mapped into this location. The other new aspect is the JNZ instruction. The program below contains both of these aspects:

```
            LDA     13          ; load AC from address 13
            LDR     14          ; load R from address 14
    start:  ADD                 ; add AC and R
            STA     15          ; send to I/O device
            LDR     12          ; load R from address 14
            IOR                 ; OR AC and R
            STA     15          ; send to I/O device
            JNZ     start       ; if AC ≠ 0, go to start
```

The program begins as in the last example by loading AC and R with the contents of addresses 13 and 14. It then adds these numbers and sends them to memory address 15, which is also mapped into the I/O device. This means that this device will receive the sum just computed and may use it as it wishes. The program then loads a new number into R, performs an inclusive OR, and writes again to address 15, sending a new quantity to the I/O device. Then, rather than continuing, the program will jump back to the first addition, assuming that AC is not 0 after the addition and OR operations.

Read pattern lput-pat3 on the CD-ROM into system Lputer to run this program. The lower-order 3 bits of memory location 15 are written into $n2$– $n0$ respectively on a single clock cycle when this address is being written to. This, in turn, is decoded to trigger one of the cells c to c', which, when active, sounds the corresponding note. The final instruction in the program at address 7 is a JNZ, which branches back to address 2. The program will continue in this loop as long as AC is not zero. Fast forward through the animation and listen to and observe the results. Also try changing the data values in memory (at locations 12, 13, and 14) to get different results. Unfortunately, unless you are running the program on an extremely fast machine, there will be a long rest between notes even at the fastest animation speed (an alternative and faster way to make music with LATTICE will be presented in Chapter 9). The current implementation of the JNZ instruction is an unconditional jump; Exercise 8.13 asks you to turn it into a conditional one.

After a laborious process, we have now constructed a computer simulation that is at least six orders of magnitude slower than the computer on which it runs, not to mention the fact that it has only 16 words of memory. Nevertheless, this simulation captures many of the essential aspects of computer design, and is not *fundamentally* different from a desktop or laptop PC. In the next and final section of this chapter, we explore some other issues that arise in modern digital computer design.

FURTHER ISSUES IN COMPUTER DESIGN

The creation of a modern computer is an incredibly complex design process, drawing, as it does, on the talents of not only digital designers, but also operating system experts, chip optimizers, manufacturing consultants, and a dizzying array of mechanical engineers for various components of the physical structure of the machine. We cannot hope to capture all of these aspects here. Nonetheless, we close this introduction to computer design with four additional topics that will prepare the student for more advanced treatments of this subject.

Microsequencing

The control unit in the L-puter consisted of a hard-wired finite state machine. There is an alternative method of implementing the same thing in "software." The quotes around this word indicate that we are not talking about a full-fledged programming language—after all, this would require another computer to run on, and then how would we construct the control unit for *this* machine within a machine? Rather, the control unit can be implemented in what is known as microcode, which runs on a micro-PC. As its name implies, microcode is a simplified type of code that contains only the microinstructions necessary to implement the control unit. The hardware on which it runs is a miniature version of a CPU, also containing just enough hardware to run this code.

Why do things this way? One reason is flexibility. With the L-puter, the control unit was simple enough that a hardware implementation suffices. However, once constructed, it cannot be changed. Microcode, like all software implementations, is much easier to change in the case of updates. It can also be adapted to multiple instruction sets. Finally, microcode makes prototyping considerably more efficient. As the control unit becomes more complex, there are more and more variations in the design process, and therefore more room for efficiency gains. Experimenting with the design of this unit by changing the microcode is one way to achieve these efficiencies. Still, a hard-wired control unit of the sort in the L-puter will almost always run faster than a microcoded one, and therefore is preferred, all things being equal.

Interrupts

I/O devices that provide input to the CPU usually have much different and slower time characteristics than the CPU itself. Take the keyboard as an example. The best typist can enter on the order of 500 characters per minute. This is considerably slower than the processor speed of the computer, and it would be a waste of time for the CPU to sit there and do nothing while these characters are being received. What is needed is some sort of mechanism to respond only when a character is typed. This is the function of interrupts, which, as its name implies, co-opts the CPU briefly in order to force it to respond to the input.

There are three general types of interrupts. The first is an external interrupt, of which the keyboard input is a typical example. Here external devices send a signal to the CPU telling it to process the given input. Internal interrupts are generated by the CPU itself and include mechanisms to switch between processes and handle exceptions (for example, unusual situations that arise such as an attempt to divide by zero). Finally, software interrupts are generated by the programmer by invoking special instructions in the instruction set included for this purpose. Regardless of the type of interrupt, there must be some mechanism to store the state of the machine, handle the interrupt itself, and then restore the state of the machine so that it can continue on its previous path. The proper use of interrupts can make the machine considerably more efficient and more responsive to external devices.

RISC and Pipelining

There are two fundamental approaches to designing instruction sets, the complex instruction set computer (CISC) and the reduced instruction set computer (RISC). CISC contains a large number of fairly specific instructions; RISC consists of a much smaller number of more general instructions (but still a larger set than in the L-puter, for example). The advantage of CISC is that each instruction has greater capability; this means that in compiling a high-level language to the instruction set, the resulting code will be shorter. The advantage of RISC is twofold. First, the smaller and reduced complexity of the instruction set means that the control unit will also be smaller. This leaves room for more registers, and therefore less register to RAM transfers will be necessary for a given task. The other advantage is pipelining. In principle, it is to carry out pipelining with CISC, but the complexity and variety of the instruction usually preclude an efficient implementation of such.

Pipelining means that more than one state in the control cycle can be executed at once. To take an example, let us say we are carrying out an ALU operation. This involves the ALU itself and the AC register. There is no reason why the next instruction cannot be fetched at this point; this involves only the AR, DR, and PC registers, and RAM and the bus. But the ALU does not need the bus. Therefore there is no conflict. This is the general idea behind pipelining. Under the right conditions, the next part of the control cycle can be executed while a previous part of the cycle is still executing. When done correctly, the situation looks like Table 8.7. As instruction one is instituting its decode cycle, instruction two begins its fetch cycle. At time step three, all three instructions are in one part of their cycle. Describing the extra hardware that will be required so that the instructions do not conflict is beyond the scope of this book. The net results of this procedure, however, is that three instructions finish in a total of five time steps instead nine.

TABLE 8.7 The Pipelining of Three Instructions

Instruction	Time 1	Time 2	Time 3	Time 4	Time 5
1	fetch	decode	execute		
2		fetch	decode	execute	
3			fetch	decode	execute

Better pipelining in conjunction with the RISC architecture is one of the key reasons that computers have continued to follow Moore's law, which states that processing speed doubles roughly every 18 months. The other reason is increased speed of the hardware on which the instruction set runs.[2]

High-Level Languages

Under most circumstances, computers are programmed in high-level languages rather than the instruction set itself. These programs are then compiled or otherwise transformed into the native instruction set of the computer. This is not a hardware issue per se, but the efficient processing of these languages will be affected by the hardware design. By far, the most important revolution in programming in the past 15 years has been the inclusion of object-oriented behavior in most new languages (e.g., Java and C++). Object-oriented languages do not contain unattached blocks of code like their non-object-oriented counterparts. Rather, all procedure and data are associated with particular classes of objects. This makes it considerably easier to read, debug, and above all, write programs. The impact of this relatively new type of program structure on hardware design, however, is still the subject of current research.

SUMMARY

This chapter builds on the foundations introduced in the previous chapter to construct a full-fledged working computational system. The first step in this construction is the design of a register set. There are two types of registers: internal and programmer-accessible. The former are used to implement the fetch and decode cycles of the fetch-decode-execute sequence and to implement the instructions. The latter are used by the instructions to store data.

Every CPU is characterized by an instruction set, and this set largely determines the design of the rest of the machine. In the case of the L-puter, we introduced a canonical set of eight instructions capable, in principle, of realizing a wide range of programs. This set included load and store instructions, a conditional jump, and four ALU instructions. The control unit is the heart of the CPU, and decides which of these instructions to execute. This finite state machine also contains states for the fetch and execute cycles. The control unit carries out these tasks by asserting the control signals appropriate to each of its states. This allows, for example, one register to read another register by placing the contents of the first register on the bus, and then asserting a load on the target register.

The final piece of the puzzle and the workhorse of the CPU is the ALU, which carries out all arithmetic and logical operations. The L-puter was constructed with two of each. The final design includes the ALU, a control unit that is clock driven, a register architecture, RAM, and memory-mapped I/O. We were able to show how this design was able to run three small programs successfully. In principle, this design with a much-augmented memory would also be able to run much more complex programs. However, the finished machine still lacks a number of components commonly found in modern machines, including among other things, an expanded register set, the ability to handle pipelining and interrupts, and secondary memory devices such as a hard disk.

EXERCISES

8.1 Why in Figure 8.2 is the input to the first XOR a logic 1?
8.2 Create a 5-bit counter by modifying Figure 8.2.
8.3 What would the operation (the rightmost column in Table 8.2) look like for each of the following instructions in the L-puter:
  (a) store register R in memory
  (b) XOR
  (c) jump if R equals 0
8.4 Justify Equations 8.1a through 8.1d from Table 8.3.
8.5 Realize the control unit given these equations (there is no need to show all feedback connections explicitly; treat them as additional inputs to the next-state logic).
8.6 For each new instruction in Exercise 8.3, list the corresponding control signals as in Table 8.4.
8.7 Show the sum of the states that assert each of the following control signals: PCBUS, ARLOAD, DRLOAD, PCINC, DRBUS, IRLOAD, ACLOAD2, ACBUS, WRITE, RLOAD, ARBUS, PCLOAD, ALU0, ALU1.

8.8 Modify Figure 8.5 by adding XOR and NAND functionality to the ALU. Assume that there is an extra selection bit ALU2 and that this bit, along with ALU0, selects among the four logical operations.

Lattice Exercises

8.9 Justify the tables corresponding to each cell in the counter system.
8.10 Add an overflow detector to the ALU system.
8.11 Write a program that adds three numbers together.
8.12 Write a program that plays a single note five times. (Hint: You will need to create a loop by augmenting AC by 1 each time through and then checking if it is 0 by subtracting 5 from it each time through.)
8.13 Change the jump in the LPuter from an unconditional to a conditional one. That is, it should only jump back to the indicated address when AC does not equal 0 (you will need to add an extra variable to the control unit that looks at the value of NZ(c), which signals this condition). Test your new program by ANDing AC with 0 right before the jump in the program in lput-pat3.
8.14 Change the AND in the ALU to an XOR. Write a program to test your change.
8.15 Write a program to play the following note sequence: c, c, g, g, a, a, g. What song is this the start of?

ENDNOTES

1. The Java programming language is actually both compiled and interpreted: it is first semi-compiled into a platform-independent set of instructions known as ByteCode, and then interpreted at runtime to meet the needs of the particular platform it is running on.
2. Another recent trend is the inclusion of multiple processors on a single chip. This is likely to be an increasing trend in the coming years as pipelining and hardware optimization reach fundamental barriers that cannot be broken.

9 Explorations in Digital Intelligence

In This Chapter

- Introduction
- Pattern Recognition
- Pattern Completion
- Inference and Expert Systems
- Neural Networks
- Learning
- Search
- Emergent Behavior
- Summary
- Exercises

INTRODUCTION

What is intelligence? This is a question that has long perplexed both cognitive psychologists and practitioners of the art of instilling computers with this quantity, those in the field of Artificial Intelligence. While there is no consensus about the answer to this question, or even what would constitute an acceptable form of an answer, two distinct strands are evident in the writings of the researchers in these fields of inquiry.

The first, and increasingly popular approach to answering this question revolves around the notion of adaptivity. Advocates of this viewpoint forego looking at the workings of the minds (or artificial brains) of the creatures (or machines) that they study and instead concentrate on objective behaviors when assessing intelligence. If the system under study is able to adapt to a rapidly changing environment, and ideally is able to pass this information to others of its species through communication or genetic propagation, then they argue that there is no reason to deny the power of this natural or artificial system's approach. Consider, for example, the ant, which has survived on this planet far longer than Homo sapiens. Or consider a (probably not too distant) future robot that can reproduce itself by scavenging for parts and energy. Whether individual members of these species are intelligent, these advocates would say, is largely beside the point: it is the behavior of the species as a whole that is clever by virtue of optimally exploiting its environment. It is not that each ant is having "important" thoughts but its cooperative behavior that has enabled it to survive both ice ages and heat waves and populate the entire planet. The self-reproducing robot may never win a Nobel Prize or create a great work of art but just might last longer than our species.

Despite the undoubted appeal of skirting the problem of intelligence by focusing on behavior and not on thought or algorithmic processes, there is a serious problem at the heart of this approach. Consider, for example, someone with the intellectual capacity of an Einstein but without his avuncular nature or social capabilities. Let us dub this virtual cousin to the 20th-century genius Altered Einstein. Altered, like Albert, has the capacity to revolutionize our understanding of both quantum mechanics and relativity. However, unlike his counterpart, Altered is completely maladaptive. He cannot cook, he cannot make his bed, he can barely dress himself, and he certainly never said anything about time passing slowly, in relativistic fashion, when sitting on a park bench with a pretty girl. In fact, Altered virtually ignores everyone, including women.

It is easy to see that a race of Altereds would quickly die out. Yet do we really want to question Altered's intellectual capacity? We may grant that Altered does not have all of Albert's charms, yet most would say that Altered ranks high in intelligence among all organisms currently alive and perhaps high among all those that have ever lived. Intelligence then, can flourish with or without the ability to adapt to one's environment, and appears to be something quite distinct from it.

This is more than merely a semantic matter; it lies at the core of what a proper approach to creating artificially intelligent machines should be about. In accord with the Albert/Altered example, we tentatively conclude that intelligence has something to do with how an organism, natural or otherwise, makes sense of the world around it (we will also bring back the adaptive approach later in the chapter, as appropriate). Specifically, we will adopt the following working definition: an organism will be intelligent to the extent that it is able to take its raw sense data, that is, the information impinging upon its sensors, and generate new and valid information on that basis.

A simple example of this is the following. Suppose you are walking down the street and you spot a friend whose face is partly obscured by a street sign. It is likely that you will have little difficulty recognizing your friend. In addition you will be able to fill in details, at least in outline, of the part of the face that is obscured. If one eye is hidden, you do not suppose that an emerald is in its place, unless you previously have seen your friend with this adornment. These types of inference are known as pattern recognition and pattern completion, respectively, and are treated in more detail below. At this point, it suffices to draw the connection between these processes and the proffered definition of intelligence. Both involve the creation of new information that was not present in the environment itself. Assuming your friend was not wearing a name tag, your recognition of him by name entails a large amount of additional information, not just the name per se, but also all the other things that you know about him to be true. Likewise, pattern completion adds to your knowledge about your friend in the absence of direct sensory evidence.

If this view has some validity, then it should also be consistent with the fact that both Albert and Altered Einstein should be considered highly intelligent. What did Einstein accomplish? He was able to formulate *general* rules for describing the dynamics of very fast particles (with special relativity), very small particles (with quantum mechanics), and revolutionized our view of the nature of gravity (with general relativity). In other words, with a few master strokes, Einstein was able to increase our informational content about every particle in the universe. Altered may have trouble negotiating a transaction at a convenience store, but he too must be considered a genius by the same token. Thus, the notion of intelligence as the potential to create new and valuable information is consistent with these paragons of intelligent behavior.

Are these esoteric matters for the emerging digital engineer? Twenty years ago, yes. Ten years ago, possibly. Today and in the coming years, definitely not. The history of digital devices is the history of devices that are becoming increasingly intelligent. Examples abound in the home from washing machines that adjust their cycles depending on the load to cameras that respond to the level of ambient light to PDAs that recognize their owner's handwriting. In business, expert systems, a type of intelligent system we will consider below, are increasingly used for such tasks as credit analysis and data mining is becoming an important component in marketing. In this chapter, we study how digital machines can add to our store of information in addition to transforming this information.

PATTERN RECOGNITION

Pattern recognition (PR) is the process of taking raw information and generating a label that provides a wrapper for that information. PR is an essential aspect of all intelligent behavior and allows, among other things, the thinking creature/machine to make sense of its world. Examples range from naming shapes to letter recognition to friend versus foe designation to face recognition. One way of thinking about PR is as information reduction (we will resolve the possible contradiction with our working definition of intelligence shortly), i.e., taking a large amount of data and reducing it to a small set of descriptions. Look at the room around you. The sheer amount of information, if described in elementary pixels, would be overwhelming, and would seriously impede any further thought process. Instead, what happens (mostly unconsciously), is that you see your world as a collection of higher-level objects: books, tables, chairs, computers, etc. This compresses your environment into a manageable collection of items.

PR is a very large topic and we will only touch on the highlights here. The major difficulty that arises with PR in the real world is that many objects that are superficially different belong in the same category. Take face recognition, for example. If a face is to be recognized properly, it must be viewed from a variety of angles under a variety of lighting conditions with a number of possible alterations present. Among these alterations can be the addition of subtraction of glasses, facial hair, hats, or other articles of clothing, and other variables such as facial expression. The ability to perform what is known as invariant recognition, or recognition in the context of these changing conditions, constitutes a difficult challenge and one that has consumed many scientist-years of research.

More formally, a working pattern recognition must attempt to achieve at least five types of invariance: translation, size, rotation, shape or form, and noise. The first three are specifically visual but the last two apply also to other modalities, such as sound. Translation invariance refers to the ability to recognize objects regardless of their position in space. Size invariance means that differences, either in absolute size or size in the visual field, will not hinder recognition. Rotation invariance implies the ability to classify an object independent of rotation either in the plane of view or in three dimensions. Shape and form invariance means that small differences in appearance or other features do not block recognition (think, e.g., of seeing a friend with a newly grown mustache). Finally, all real data will be noisy to some extent, either because parts of the object will be occluded or because of errors introduced in processing.

PR is often treated as an end in itself, which is understandable given the difficulty of the task. But the real reason for doing PR, and one that is consistent with our working definition of intelligence as information addition, is that categorization allows additional inferences to be made about the object. Suppose again that you are walking down the street and you spot a friend that you recognize. A host of information is immediately at your disposal, information that would not be present if you simply saw your acquaintance as an unprocessed set of color patterns. You know your friend's preferences, his manner of speech, his history, perhaps his annoying tendencies, all of which allow you to predict his actions before they actually happen. Note also that your placing a label on the object (my friend Bill) does not preclude your ability to register lower-level changes. If Bill has that new mustache, you still see it in addition to recognizing his countenance. Thus, information is not necessarily lost when doing PR, but much valuable information can be gained.

In summary, PR is an essential aspect of first reducing the world to its constituent parts, and then generating new information about the world on the basis of that reduction.[1] Without such a step, it is generally not possible to form reasonable predictions about the course of future events in a sufficiently efficient manner.

From what has been said so far, it should be apparent that building a competent PR system is no easy task. Nevertheless, many of the basic principles can be illustrated with the LATTICE system. Loading pr1 from the Chapter 9 folder on the CD-ROM illustrates the most basic type of PR. This system recognizes two block letters, "A" and "C," with the labeled "letter" machine moving to the appropriate state when these items are present. As Figure 9.1 illustrates, the system also is noise tolerant to some extent. This is the result of not requiring every pixel to be present, as can readily be seen by looking at the lower bound parameters in the definition of the variables.

Recall that one of the goals of a PR system is to remain responsive as the shape of the target pattern changes. System pr1 does this to some extent, as Figure 9.1 shows, but will not work if the patterns are stretched in the vertical or horizontal directions. The problem is that the system is expecting a fairly specific representation for each letter. One way of circumventing this is to perform PR in stages, first recognizing features and then building from there. For example, a block "C" is composed of a vertical line with a horizontal line attached at the top and bottom; the exact length of these is not critical. System pr2 takes such an approach to letter

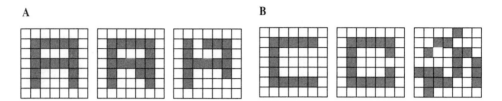

FIGURE 9.1 (A) Some patterns recognized as "A"s, and (B) as "C"s by the PR1 system.

recognition. In the first stage, three features are extracted corresponding to the patterns at the top of Figure 9.2. These features detect the junction of vertical and horizontal bars. Advancing the simulator one step yields these junctions, and in the next advancement the system detects whether an "A" or "C" is present, depending on the location of these junctions. By providing some latitude in the placement of the features, stretching of the letters can be handled, as shown in Figures 9.2(B) and 9.2(C).

This second PR system still does not provide all the invariance we might need. For example, it does not handle rotated letters. The reduction of an object to constituent features, however, is the first step in providing this and other forms of invariance. A full-fledged system would require a mechanism that detected invariance in the relation between extracted features in order to achieve this goal.

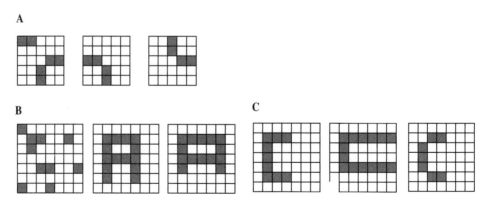

FIGURE 9.2 (A) The initial features for detecting junctions in system PR2, (B) some stretched "A"s recognized by the system, and (C) some stretched "C"s recognized by the system.

PATTERN COMPLETION

The goal of pattern completion (PC), unlike pattern recognition (PR), is not to classify the pattern but to restore it to a canonical state. There are two ways to accomplish this. One is to first do PR and then have the resulting category turn on features that were absent in the original pattern but that should have been present, and turn off features that were present but should have been absent. For example, if part of a letter was missing, as long as it could still be classified, this classification could be used to restore the missing area. The problem with this method is that it assumes that the PR is successful, which is not always the case. Take again the example of someone walking down the street, only this time it is a stranger. You are on the opposite side of the street walking in the other direction, and you only see the left half of his face. It is not necessary to know him to infer that the features on the right side are likely to match those on the right. Moreover, this PC could not be the result of first putting him in the general category of person (as opposed to a specific person), because your filling in will depend on what you see on the exposed part of the face.

Thus PC is an essential component in the repertoire of an intelligent system. As with PR, it can work in two general ways. The first method is to aim to restore a pattern as a whole. For example, the target input could be a canonical set of block characters. The goal of the system would be to restore a possibly distorted pattern, or one with pixels absent, to the closest of these characters. The system may also decline to restore the pattern if it is so degraded that it resembles none of them. This is traditionally done with neural networks, a type of intelligent system described in more detail below.

The other method is to restore individual features such as lines or other higher-level collections of elementary inputs. This provides somewhat greater flexibility than the character restoration method in that the precise target need not be known in advance. For example, suppose the characters are those in a foreign language that you are not familiar with, or even an English character in an unknown font. The fact that a small number of pixels are absent or shifted should not prevent restoration of the character to a canonical form. However PC is accomplished, the consistency with our working definition of intelligence should be immediately apparent. Information not present (or wrong) in the original pattern is corrected by the working system, providing the system with more informational content than was originally present in its environment.

System pc1 in the Chapter9 folder on the CD-ROM provides an example of character restoration, in this case to the desired "X" shape. Figure 9.3(A) shows some examples of the diversity of the patterns that can be restored. Depending on the original pattern, the system may take more than one step to reach what is known as a fixed point, that is, a pattern that no longer changes. PC1 works with two relatively simple rules, one that attempts to restore the white space in the "X," and one that attempts to restore the diagonals, as examination of the variable definitions of I0 and I1 and the state table reveals.

System pc2 is a system that tries to restore incomplete vertical and horizontal lines. Figure 9.3(B) illustrates the versatility of this system, which works on a variety of patterns. As examination of the definitions of I0 and I1 reveals, this system has no real knowledge of letters per se. It simply looks for constituents of lines and then fills in the gaps if there are any. This system would need to be augmented by additional mechanisms to restore white space between these lines should this space have extra pixels activated.

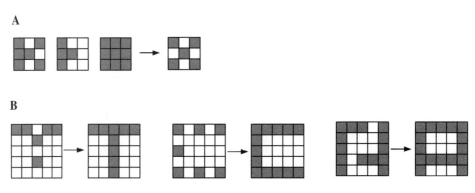

FIGURE 9.3 (A) Example of PC1 restoring an "X," and (B) examples of line restoration in system PC2.

INFERENCE AND EXPERT SYSTEMS

Inference may properly be deemed the pinnacle of intelligent mechanisms, and for good reason. Inference encompasses any mechanism, whereby the intelligent system adds to its database of knowledge by drawing implications from the current state of this database and the current state of the world. Needless to say, this coincides perfectly with our working definition of intelligence. Once again, we see that an intelligent system is one that is not limited to the immediate sensory content of its environment but also augments this content with additional information.

TABLE 9.1 The Truth Table for the Implication Connective

P	Q	$P \Rightarrow Q$
0	0	1
0	1	1
1	0	0
1	1	1

Although inference is defined in these general terms, the most prominent type of inferential mechanism involves what are known as Horn clause rules. Horn clause rules are characterized by a one or more conjunctive conditions followed by a conclusion that follows if these conditions are met. More formally, we may represent a Horn clause rule as

$$(P1 \cdot P2 \cdot P3 \ldots \cdot PN) \Rightarrow C, \tag{9.1}$$

where the truth table for implication connective, $\Rightarrow$, is shown in Table 9.1. Note that $P \Rightarrow Q$ is an alternative way of writing $P' + Q$ (see Exercise 9.4).

As the table shows, suppose we are given that $P \Rightarrow Q$ (read "P implies Q"), that is, that this condition is 1, and that we also know that P is 1. The only row in the table in which both of these conditions hold is the last row. But this row shows that Q is 1. Thus, if we know P and we know that $P \Rightarrow Q$, we can conclude Q. This rule of inference is known as Modus Ponens, and forms the basis for rule-based systems such as expert systems.

As an example, consider the following hypothetical rule, for approving a car loan, written in the preferred vertical format:

(R1)   IF   credit rating is high, and
debt is low, and
income is high, and
employed, and
own home

THEN   approve loan

The rule states that if all five of the conditions above the line are met, then the conclusion is also met, which in this case means that the loan will be approved.

In practice, strict Horn clause rules are inadequate to express the fuzzier types of inference we might wish to do in the real world. For example, in the above rule, do we really want to require every condition to be true? No doubt every buyer that meets these collective conditions will pay back his loan, but we will lose many otherwise qualified buyers by making our conditions so stringent. One way of relaxing the conditions is to allow some leeway as to the number of conditions which must be met. This is illustrated with these two rules:

(R2) IF   credit rating is high, and   (R3) IF   credit rating is high, and
            debt is low, and   debt is low, and
            income is high, and   income is high, and
            employed, and   employed, and
            own home (4 or more)   own home (exactly 3)
      THEN approve loan   THEN consider loan

Rule (R2) states if four or more of the conditions are met, then the loan is still approved. For example, if someone has a good credit rating, has no or little debt, has a high income, and is employed, it probably doesn't matter if they own their home to receive a car loan. Rule (R3) states that if three of the five conditions are met, you might still want to consider granting the loan, or altering it by requiring a larger down payment, for example.

Reducing the number of conditions that need to be met gives us more leeway, but is usually still not sufficient to capture the fuzziness of real world conditions. For example, it would be nice if the applicant's income was high, but this excludes middle-class buyers. What is also needed is a way of making individual conditions fuzzier in addition to the rule as a whole. This is shown in the following modification of (R2):

(R2') IF   credit rating is good or high, and
             debt is low or moderate, and
             income is average to high, and
             employed, and
             own home (4 or more)
       THEN approve loan

Each of the first three conditions have been widened to include other possibilities. For example, it is no longer necessary to have a high income; anything in the range of average-to-high will do. This is a threshold scheme in that a condition is allowed to influence the final decision if it meets or exceeds a given criterion.

There are a number of other numeric schemes in which conditions are rated to the extent to which they are met, and then these ratings are combined to yield a score that indicates the extent to which the conclusion holds. Such schemes are beyond the scope of this section. One additional mechanism that we *will* consider is rule chaining. Typically, expert systems are not flat as in (R1) and (R2) but have additional hierarchical rules, which feed into the conditions of other rules. For example, while it is possible to obtain a generic credit rating from a credit agency, it is often desirable to customize this to the given context. Rules (R3) and (R4) represent such a possibility for the first condition in (R2′):

(R4)	IF	no bankruptcy, and no missed payments, and borrowing moderate (3 of 3)	(R5)	IF	no bankruptcy, and no missed payments, and borrowing moderate, (2 of 3)
	THEN	credit rating is high		THEN	credit rating is good

Rule (R4) states that if all three of the antecedent conditions are met, then for the current purposes, the credit rating will be considered high, and (R3) states that if at least two of the conditions are met, then the rating will be good. These rules will be evaluated first, and the result will then influence rules (R2) and (R3).

There are other practical aspects of rule-based systems that we will just mention here. First, there is the issue of what to do if more than one rule is triggered at the same time. Next, there is the issue of what is known as control, that is, the order in which to gather data from the user of the system. Finally, there is the issue of explanation: a system that merely comes up with a solution but cannot explain its reasoning may be unsatisfactory from the point of view of human users. Each of these must be fully addressed in the case of true working system.

ON THE CD

The following LATTICE-based systems parallel the development from the simple to more complex expert systems in this section. System rule1 on the CD-ROM signals that the loan is approved only if all five of the antecedent conditions are met (pattern files rule1-pat1 and rule1-pat2 cause the loan rule to fire and not fire, respectively). System rule2 mirrors rules (R2) and (R3) in that the former fires if four or more conditions are met, and the latter if exactly three are met (this can be seen in the definitions of variables I0 and I1 for the "loan" cell). Patterns rule2-pat1, rule2-pat2, and rule2-pat3 cause the loan to be granted, to be considered, and to be denied respectively. System rule3 is identical to system rule2 with the exception that additional cells feed the first three conditions. Only if the top two possibilities are present will the condition meet the threshold and be propagated to the cells to the right. Patterns rule3-pat1, rule3-pat2, and rule3-pat3 cause the loan to be granted, to be considered, and to be denied respectively. Finally, system rule4 is identical to rule3 except that the first condition is fed by additional conditions, mirroring rules (R4) and (R5). If two of the antecedent conditions are met, then it is concluded that the credit rating is good and if all are met then the credit rating is high. The behavior of this system is otherwise identical to system rule3, with patterns rule4-pat1, rule4-pat2, and rule4-pat3 causing the loan to be granted, to be considered, and to be denied respectively.

While in theory there is nothing preventing one from implementing expert and other rule-based systems in hardware (see e.g., Exercise 9.4), in practice they are usually implemented in software. The reason for this is threefold. First, the size of the rule base is usually quite large, comprising anywhere from hundreds to thousands of rules. Such a system would be awkward, although not impossible, to realize in hardware. Next, most expert systems shells (general frameworks for implementing expert systems in specific domains) allow one to call general procedures to compute whether rule conditions are fulfilled. This significantly expands the capability of a system by allowing, for example, arbitrary numeric computations. Finally, and most importantly, the rule base of an expert system is usually in a constant state of flux, and therefore a software implementation is much more appropriate. Hardware implementations, while certainly conceivable, are best reserved for small and fixed systems for which computationally speed is of the essence.

NEURAL NETWORKS

If we look at the brain, we see nothing like the high-level rules of the previous section. Instead we see a large number of neurons, approximately 10^{10}, connected in an intricate and complex web. It is unlikely that any individual neuron represents a coherent concept; otherwise if it was lost (a not uncommon occurrence), than that concept would be lost also. In other words, the brain is thought not to contain so-called "grandmother" cells, or cells representing abstract concepts such as one's grandmother. Instead, the notion of one's grandmother and all other higher-level concepts are believed to emerge as a result of the collective action of many thousands of cells.

In recent years, this notion has been transferred from the study of the brain per se to the arena of intelligent computing. The attraction to neural computing is twofold: first, there are 6.5 or so billion existence proofs that intelligence can be engendered in this fashion, namely every living human being. Second, and more theoretically, the direct brute force approach of manipulating high-level knowledge, such as with an expert system, may be adequate in certain domains but is likely to be inadequate in describing the exquisitely precise and rapid processing necessary to respond to a rapidly changing real-time environment. To take just a few examples that people excel at and in which robots are currently still somewhat clumsy: the combination of phonemic, grammatical, and semantic processing capable of extracting meaning from speech almost as it is uttered, the likewise near instantaneous processing of the visual field, and the ability to plan complex movement involving many muscles with little or no effort. All of these and other domains are probably best viewed as the result of the collective action of parallel, but simple algorithmic components.

The central such component in neural computing is the model neuron, or a neuron from which all organic incidentals have been stripped away, such as the blood supply feeding the neuron and the ions necessary for its proper behavior, to reveal its underlying algorithmic essence. This essence is believed to be a threshold system. More specifically, the model neuron sums the product of the firing rates and the synaptic efficacies, or weights, that lead into it, and if this sum exceeds a threshold then the neuron fires. That is, if

$$\Sigma \, a_i \, w_{ij} > \theta, \qquad (9.2)$$

where a_i is the firing rate of the ith neuron, w_{ij} is the weight between the ith neuron and the jth neuron (the neuron under consideration), and θ is the threshold, then the jth neuron fires. Figure 9.4 illustrates how this calculation works in a particular case. The three bottom units are firing at relative rates of 0.75, 0.5, and 0.25, and are connected to the top unit with weights of 1.0, 0.5, and −0.5, respectively. A positive weight is excitatory and a negative weight is inhibitory. The net activation for the top neuron is therefore 0.75 * 1.0 + 0.5 * 0.5 + 0.25 * (−0.5) = 0.875, which is greater than the threshold q of 0.75, and therefore the top neuron fires. In practice, the firing rate is usually computed as a monotonically increasing nonlinear function of the difference between the net activation and the threshold, rather than a strict threshold function, although we will ignore this aspect here.

A network of such units is constructed in order to solve a particular task or to simulate a particular phenomenon. Figure 9.5 presents one of the earliest illustrations of neurally inspired simulations. The Necker cube on the left of the figure is one of the most famous visual patterns in psychology and it illustrates, among other things, that vision is not merely a matter of registering pixels but of constructing an image. In this case, the construction is of a 3D cube but there are two

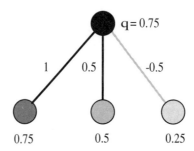

FIGURE 9.4 A small network in which the three bottom units feed the top unit with indicated weights.

possible views, one with the a-b-c-d face in front and one with the e-f-g-h face in front. Concentrating on one of these corners can bring its face to the forefront, flipping one's view of the cube.

The neural model in Figure 9.5(B), in which activity in a unit represents the fact that this corner is seen in the forefront, captures this effect. Each of the corner units is connected in an excitatory fashion with the other corners in its group. In addition, there is an inhibitory connection between every corner in one group with every corner in the other group (in the figure, this is represented as a single large bidirectional inhibitory connection). The net result of this model is that a small amount of priming activation to any of the corner units, representing attentional focus on that corner, will cause its group to become fully active. In addition, when one group becomes active, it inhibits the other group. Thus, the model effectively represents two features of the cube: (1) only one group is at the forefront at any given time, and (2) attentional focus can flip the groups.

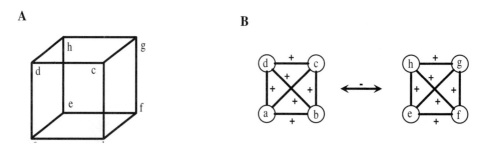

FIGURE 9.5 (A) The ambiguous Necker cube, and (B) a neural model of the ambiguity.

ON THE CD

Neural networks are continuous analog systems because the weights and activations are real numbers. Nevertheless, it is possible in this case to produce a reasonable digital facsimile of the Necker cube model. System net1 on the CD-ROM contains two partitions in addition to the background partition, representing the two views of the cube. Variable I0 in each partition encodes the excitatory aspect of the network, variable I1 encodes the inhibitory aspect, and variable I2 encodes the priming aspect. Priming occurs when a cell around one of the face cells is active; this can be seen by loading pattern net1-pat1. This first causes the corresponding face cell to become active, which in turn triggers the other units in the face (in the next simulation step). Pattern net1-pat2 shows what happens when both faces are primed unequally. The first face receives greater initial priming, and therefore ends up winning the competition between the views. Finally, net1-pat3 shows what happens when both receive equal priming; neither view is favored.

One of the nice properties of neural networks is that they provide a clean mechanism of combining what is known as bottom-up and top-down reasoning. Bottom-up processing refers to reasoning from sensory or other input data, and top-down means reasoning from expectations or predispositions. There are many seemingly intractable chicken-and-egg-type problems that arise in intelligent computing which purport to show that neither mechanism could possibly suffice to solve the problem at hand. For example, the so-called cocktail party effect refers to the ability to recognize your name being spoken from across the room in a crowded party. If you were paying attention to all the noise around you (bottom-up processing), just in case your name was mentioned, you would be devoting attention to an overwhelming amount of information. On the other hand, if you concentrated on your name or other salient features about yourself only (top-down processing), you would be unduly distracted from your current conversation. So just how is it possible to both pay and not pay attention to egocentric triggers in your environment?

One resolution of this dilemma is to have such egocentric features be partially active.[2] If such a feature then receives additional and possibly subliminal (i.e., unconscious) activation from sensory data, it exceeds threshold and attention shifts in that direction. If no such activity is received, these features are ignored. In other words, the cocktail party effect is the result of partially active top-down features and low-level triggers in the environment becoming engaged in a bidirectional resonance in which both become reinforced via mutual excitatory connections.

Figure 9.6 illustrates another similar situation. Here part of the first letter of each word is obscured, and by the available bottom-up (i.e., sensory) evidence the letter could be either a P or an R. However, preexisting top-down information regarding possible words allows one to disambiguate the situation. Because RIN is not a valid English word, the first case is seen as PIN, and because PIM is not valid, the second is seen as RIM. In this situation, only cooperation between bottom-up and top-down data yield the best guess as the actual content of sensory information.

FIGURE 9.6 The first word is seen as PIN and the second as RIM despite the fact that the first letter is identical in both cases.

The disambiguation of sensory data by top-down processing is an exemplary instance of how an intelligent system augments information. Preformed expectations help not only in this sense, but also in anticipating future action and thereby reducing reaction time. Without this skill, it would be almost impossible to hit a fast ball in baseball or return a well-struck serve in tennis. Conversely, violating your opponent's expectations, such as in tennis, "wrong-footing" the opponent by hitting the ball in the direction they least expect, often wins the point immediately.

ON THE CD

System net2 on the CD-ROM contains a simulation of the just described problem. Each letter is represented in a 5 by 5 grid, and each contains an internal pattern-recognition cell, the contents of which are transferred to an external recognition cell above. The latter are designed to indicate whether there is both bottom-up and top-down confirmation of the presence of the letter. Advancing the simulation once causes all letters to be recognized, as shown by the newly active purple cells. On the next advance, these recognitions are transferred to their respective copies. Note that the system is considering both P and R as possibilities for the first letter. On the next advance, the word PIN is triggered; as can be seen by examining variable I0 for this partition, this occurs only if all three letters are present. The final advance turns off the R because it is not consistent with this word. Thus, only P remains active because it has both top-down and bottom-up support. In Exercise 9.5, you will be asked to augment this system so that it works for RIM also.

Visual processing is another area in which neural networks excel. Recall from the section at the top of this chapter on pattern recognition the importance of segmentation of a complex environment into distinct objects. In the human mind, this is both a categorical *and* perceptual effect, with the latter aiding the former in its attempt to divide up the world into distinct chunks. The first and most crucial step in this process is the enhancement of lines in the visual image. Figure 9.7 provides a metaphorical view of this process. In the actual image, one part of the image blends into another without clear distinction (panel A). However, early processing extracts the lines in the image (panel B), and when these are added to the original image (panel C) the distinctions between objects becomes more evident. This is why it is almost impossible to look at the world and perceive a sea of colors only; by the time the image arrives at the conscious mind the edges between objects have already been highlighted.

FIGURE 9.7 (A) Actual pixels, (B) after line extraction, and (C) what the mind's eye sees.

Line and edge extraction fits well within the neural network paradigm because it is best explained as the application of filters to each pixel. That is, rather than consisting of a sequential run through a single complex algorithm, as in conventional Artificial Intelligence, early visual processing consists of large numbers of parallel, but simple processes. Figure 9.8 presents three such filters, one for extracting vertical lines, and two for left and right edges respectively. The filter in panel A is shown as a neural network in panel D. The weights in the grid correspond to the connections in this network. When a row of three pixels is present, as in this drawing, then the "line present" node will receive superthreshold activity, assuming that an on pixel triggers an activation of 1 (the total net activity will be $1*1 + 1*1 + 1*1 + 0*(-1) + 0*(-1) + 0*(-1) + 0*(-1) + 0*(-1) + 0*(-1) = 3$). If however, not all three vertical units are active, or if a unit to the side of the line is active, then the total activity will not achieve threshold for the detector unit. The edge filters work similarly. In these cases however, it is necessary for pixels to be active to the side of the edge (an edge, by definition, is a point of contrast between dark and light).

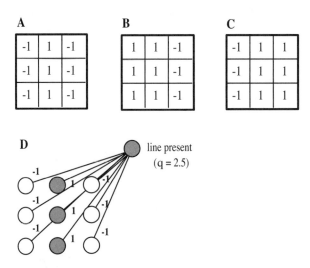

FIGURE 9.8 (A) A line filter, (B) a left edge filter, (C) a right edge filter, and (D) a network for a single pixel corresponding to the line filter.

The parallel aspect of this process resides in the fact that there is one such filter for every pixel in the image. A new filtered image is thus rapidly constructed by applying these filters simultaneously to yield an image as in Figure 9.7(B). In addition, in a real network, and most probably the brain also, the degree of lineness or edgeness can be obtained when pixel values are continuous between 0 and 1, rather than discrete binary values, by noting the degree to which the detector exceeds threshold.

Although it is difficult to handle continuous values with digital systems such as LATTICE, we can nevertheless illustrate the essence of filtering with some degree of ease because LATTICE is an inherently parallel system. System net3 on the

ON THE CD

CD-ROM extracts out vertical lines by applying a logical filter to each cell. This filter only allows those cells to remain active whose top and bottom neighbors are also active but have no neighbors to the side active. System net4 works similarly to extract vertical edges as applied to these patterns. Exercise 9.7 asks you to add horizontal edge filters to the logical expression in this system. The extracted image should now be closer to the kind of line drawing in Figure 9.7(B); the addition of diagonal edge filters would bring it further in this direction.

If neural networks were merely a more robust way of achieving pattern recognition, they would retain some degree of interest, but would not generate nearly the amount of excitement that they have. The additional factor driving much of the attention toward these models is the fact that they, like the brain itself, are capable of learning to adapt to their environment. The next section provides a general framework for exploring the learning problem. Here we will briefly explore a key moment in the history of neural networks as it relates to learning and adaptation.

In the early 1960s, a researcher named Rosenblatt conceived of what he believed to be a general learning method. He termed his invention the Perceptron, and showed how it could achieve the kind of pattern recognition that we have been exploring with the appropriate training technique. Rosenblatt believed that a network, instilled with his learning algorithm and placed into a sufficiently complex environment, might gradually build up to an intelligence on par with humans. However, his hopes were dashed in the late 1960s when two other researchers, Minsky and Papert, showed that there were severe limitations on the kinds of things the Perceptron could represent, and therefore learn.

Figure 9.9 shows the paradigmatic example of the Percetpron's limitations. As it turns out, the Perceptron is only capable of dividing up the example space with lines (or hyperplanes in more than two dimensions). In order to work, all true (or digital 1) instances of a function must be on one side of the line and all false (or digital 0) instances must be on the other. But this is impossible with the XOR problem, which as we recall from previous chapters, is 1 if and only if P1 and P2 are not equal. As can easily be seen in the diagram, there is no line that places these examples on one side of the line and the others on the opposite side. For example, the thick line in the figure has (0,0) on one side but the remaining three examples together.

Fortunately, another neural network learning algorithm known as backpropagation eventually came to the rescue. Backprop, as it is colloquially known, was able to learn the XOR problem and many other similarly complex problems with the help of a hidden layer, or extra layer of units between the input and output decision layer. This advance helped spur the rebirth of neural networks, and they now

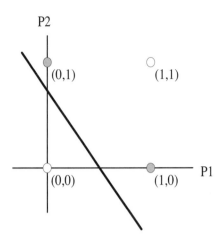

FIGURE 9.9 No dividing line can place the shaded circles and the clear circles on opposite sides.

stand as equal partners in Artificial Intelligence with models using more traditional representations. There is one gnawing problem that has yet to be fully addressed, however. Not only is there no evidence that the human brain performs something like backprop, there are strong hints that it is unlikely that it is capable of implementing this algorithm. But we also know that people are exemplary learning machines. One way out of this conundrum is to postulate that people use a higher-level learning method such as that presented next.

ON THE CD

With LATTICE we could, of course, cheat, and implement XOR in single level by choosing this operator directly, or by forming the equivalent expression P1P2′ + P1′P2. The challenge, however, is to do this in a manner similar to the Perceptron, which is equivalent to saying that simple sums or products are allowed but not arbitrary SOP functions. System net5 on the CD-ROM shows one way of implementing XOR within these constraints with a three layer system, similar in the manner in which backprop solves the same problem. The right cell in the middle layer is active only when both of the input cells are active. The left cell in the middle layer is active when at least one of the inputs is active. The top recognition unit responds when the either cell is active and the both cell is inactive. As can be seen by applying the various combinations of inputs to P1 and P2, the system makes the correct classification in all cases after two simulation steps.

LEARNING

All of the mechanisms previously described, as has been argued, involve some means of squeezing more information out of what is originally provided to the system. But how does this system gain the pieces of knowledge that act as the basis for these mechanisms? This can be done directly, one by one, but this is awkward as well as time consuming. It would be far better if the system could acquire its knowledge by its own accord. For example, instead of describing to a system what it means to be the letter P, and all the possible variations on this letter, it would be nice to simply provide a few representative examples. Ideally, the system would then be able to generalize from these examples to other instances which it has not yet seen.

This paradigm constitutes our first example of learning, and is known as supervised learning. As the name implies, supervised learning consists of a teacher telling the system how to classify patterns in its environment. The teacher need not, however, provide every possible instance; it is up to the learner to fill in the missing details. Our old friend the Karnaugh map provides a convenient way of viewing this process. Let us assume that we are trying to teach you about an object you have never encountered, which we will call an "xygraby." We provide you with two instances of things that are xygrabies, and two instances of non-xygrabies, each of which is characterized by four features. The map in Figure 9.10(A) graphs these instances. The "+" marks indicate positive instances of the concept and "?" marks negative instances. The first xygraby is red, small, round, and shiny, and the other examples can be described accordingly.

The supervised learning task can be described as follows, given this layout. The task is to draw one or more rectangles each containing 2^n cells such that the rectangle covers all of the positive examples and none of the negative examples, or noninstances. This is very similar to the original problem of finding prime impli-

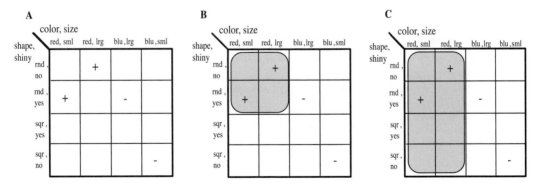

FIGURE 9.10 (A) A supervised learning problem on a Karnaugh map, (B) one narrow solution, and (C) a more general solution.

cants, except that the rectangles are permitted to cover empty space, but prohibited from covering the negative instances. Unlike the prime implicant problem, this task, by necessity, is ambiguous. In general, there will be many ways to achieve this goal, and it is this fact that makes the supervised learning problem more difficult, but also more interesting than the minimal sum problem.

Figure 9.10 illustrates two such ways. The first narrowly covers the two positive examples and the second covers these but also more empty space below. Which is correct? The answer is neither, as it depends on the circumstance. The first coverage is less risky in the sense that it generalizes less to unknown examples (the empty spaces). On the other hand, by being risk-averse, this method may misclassify examples of the target concept that are genuine. In this particular example, the narrow solution states that an xygraby is red and round. The riskier solution states that an xygraby is simply any object in the universe of discourse that is red. The former is less likely to classify a non-xygraby as an xygraby, but is more likely to miss objects that are really xygrabies.

At the risk of complicating the situation further, it is worth mentioning that there are instances in which it is justified to cover negative as well as positive examples. For example, suppose that a lone negative example is swimming in a sea of positive examples. This negative example, if taken seriously, might end up fragmenting the description of the positive examples. It may be better under these circumstances to throw out the negative example by concluding that it is an erroneous piece of data and thereby preserving the simplicity of the classification.[3] Exercise 9.8 asks you to explore this possibility in LATTICE.

ON THE CD

System learn1 on the CD-ROM contains a generalization rule for forming a 2 by 4 rectangle whenever at least two of the cells in the rectangle are positive examples and there are no negative examples. Applying this to the default configuration yields the result in Figure 9.10(C). Note that adding a negative example to the bottom-left corner will prevent this generalization. Note also that a comparable horizontal generalization rule will fail because of the blue, large, round, and shiny negative example. Exercise 9.7 asks you to confirm this.

In supervised learning, the system is provided with examples that are labeled as positive or negative instances of the target class. In contrast, in unsupervised learning, there are no such indications. All examples are unlabeled, and it is the job of the system to divide them into representative classes. The hope is that these divisions will yield information about a member of a particular class simply by virtue of being a member of that class. Suppose for example, that a marketer is trying to understand his customer base. To treat each customer as a unique instance yields little in the way of information. However, by dividing up the base into similar demographics, he is able to make sense of his customers and predict their habits. For example, a typical demographic might be young married women with small children. He knows that this is a significant group in that it has similar buying habits both obvious (diapers) and perhaps nonobvious (novels by a certain author). Not every member of the demographic must share every habit of the group, of course, but the class is sufficiently coherent that carving up the world in this way tells the marketer more about it than if he were to look at it as an undifferentiated collection of individuals.

One means of achieving this type of grouping is to cluster examples into the same group if their mutual distances in feature space is relatively small. That is, if the examples share a significant number of features, or a set of features that have a priori been deemed important, then they are put in the same class. Figure 9.11 il-

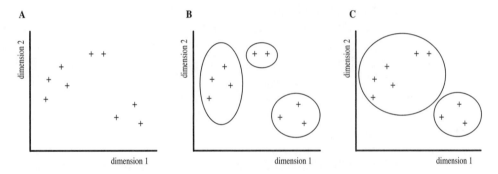

FIGURE 9.11 (A) An unsupervised learning problem graphed on two continuous dimensions, (B) one solution, and (C) one solution with larger and fewer groupings.

lustrates this process. A set of examples are graphed on two axes representing two continuous dimensions (we could also use a larger number of discrete or continuous dimensions but this would be harder to visualize). All examples are positive and are treated equally. The task is to form significant groupings, based on similarity, which in this mapping is simply physical distance. Figure 9.11(B) shows one such grouping. The original set of examples has been divided into three similar groups. However, as in the case of supervised learning, there is no one right solution. Under differing circumstances, the less strict grouping in Figure 9.11(C) may be more appropriate. Other factors and perhaps later testing of the utility of the categories will be necessary to differentiate between these two cases.

Systems learn2, learn3, and learn4 on the CD-ROM apply progressively more liberal classification criteria by weakening the conditions for forming new classes. Learn2 requires that there are at least two examples in a square of length 3 centered on the current cell, learn3 requires that there are three examples in a square of length 7, and learn4 requires that there are at least two examples in a square of length 7. Try applying these three systems to the default configuration to see how they affect the size and coherence of the formed groupings.

We now briefly explore one additional type of learning that has in recent years become more important as researchers attempt to create biologically inspired systems. These are genetic algorithms in which the learning method reflects the evolutionary process. Evolution has no explicit knowledge about either organisms or their environment. Nevertheless, it is able with remarkable precision to tailor the former to the latter, or in certain cases such as Homo sapiens, enable more general adaptive mechanisms so that the organism can thrive in a wide variety of environments. It does this through survival of the fittest.[4] Only those organisms that are fit will thrive, and only these get to pass on their genes to their children. In this way, organisms that are not ideal for the given environment will eventually die out, to be replaced by those that thrive under these conditions.

Genetic algorithms work along the same lines. Figure 9.12 provides a schematic for this process. The genes of two parents are combined, with a mixing algorithm, to produce the genes of the child. This process can also be affected by random changes in the resulting set of genes, which are known as mutations. Mutations are usually maladaptive, but on occasion they allow for the genetic pool to move in a new and interesting direction. The result of this mating process is the child, which contains elements of both parents and possibly some new elements. This is known as the genotype of the organism. This genotype then gets translated into the behavior and appearance of the child, which is known as the phenotype of the organism. If this object thrives in the given environment, then it is able in turn to mate and have its genes affect the next generation (this is shown in Figure 9.12 by the line labeled "reproduce" indicating that the child becomes a parent itself). If it dies out, this does not happen. In this manner, over many generations, the gene pool adapts to the given environment.

This process is extremely general and need not be applied to the literal creation of organisms. For example, one of the most common uses for genetic algorithms is in complex search processes. Search, as we will learn in the next section, is one of the essential aspects of intelligent systems and entails looking through a space of

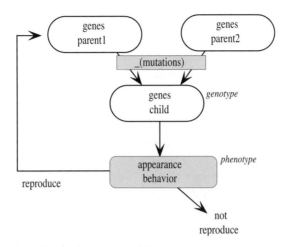

FIGURE 9.12 A schematic of the genetic process.

possibilities to find a point or set of points that maximizes a formula known as the fitness function. Different sets of genes occupy different locations in the search space, and those that produce relatively high values for the fitness function, as determined by the performance of the phenotype of a given set, are allowed to reproduce. By this process, after a sufficient number of generations, the gene pool will encode genotypes whose phenotypes satisfy the fitness function better than the original pool. Genetic algorithms work best when the problem is complex and there is no clear analytic solution, as in the following LATTICE example.

ON THE CD

System learn5 on the CD-ROM shows a square environment which is ringed by four binary genes. These genes influence the square, which in turn influences the note and rhythm cells, with the top and left genes having more influence on the note partition and the lower and right genes having more influence on the rhythm partition. Thus, the genes code for pieces of music. Try playing with different combinations of genes to change the music. Like genetic combination in the animal kingdom, however, the phenotype is not a simple addition of the contributions but a complex nonlinear function of the two.

SEARCH

Search techniques are predicated on the idea that the information an intelligent system needs is not directly at hand, even in a degraded form. However, it is implicit in the knowledge base of the system, and by exploring the possible implications of this base, it may be able to be found. Perhaps the most common type of search that humans perform involves planning. It is usually not sufficient, even in the simplest of circumstances, to take the most direct route from one's current state to a particular goal. Rather, one must examine the consequences of each move, and replan accordingly as circumstances dictate.

For example, suppose you need or desire $20,000. Let us further suppose that there is an old-fashioned bank near you—the kind where you are not separated from the teller by 4-inch bulletproof Plexiglas. The easiest way to get the cash into your pocket is to stroll into the bank with an appropriate weapon and demand the money. Putting aside moral considerations, it is not difficult to see that this direct route is probably too risky. Police departments have the annoying tendency of taking bank robberies very seriously, and your chances of keeping the money for two weeks let alone two hours is likely to be very slim. You need to search for an alternative plan, one that is less likely to yield the cash as quickly, but one that will allow you to hold on to it for a longer period of time.

Another example is route planning. The shortest route, as close as possible to the route the crow flies, may be initially appealing, but there are many potential obstacles—for example, whether the route has heavy traffic or whether there is construction—that must be considered. The key in this case, as in all planning, it to envision the consequences of an action without actually carrying it out, and then weighing its merit relative to other possibilities.

ON THE CD

System search1 on the CD-ROM provides a simple example where taking the "long-cut" proves worthwhile. There are two routes from the start cell on the left to the final cell on the right. The topmost route seems to be the shortest, but running the simulation proves this wrong. This route involves a considerable delay and the superficially longer route proves the swiftest.

LATTICE, by its nature, is capable of exploring both possibilities in parallel in the previous example. Search, however, is usually implemented on serial machines, with each branch explored sequentially. This is best visualized as a search tree, such as that shown in Figure 9.13(A). It is easy to see in this case that taking the right path on the tree leads to a shorter total time.

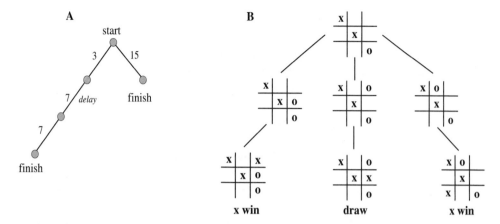

FIGURE 9.13 (A) A search tree with two paths, and (B) a search tree for tic-tac-toe.

One of the earliest uses of search was in game playing, and in fact, the best of the computer chess machines, which now stand very close to the strongest human player, still rely heavily on search-based techniques. Figure 9.13(B) shows a position, at the top of the tree, of a much simpler game, tic-tac-toe. It is o's move, and he has six total choices. Three of these, however, are symmetrical to the other three so it is sufficient to consider one half of the possibilities. If o makes the move on the left branch of the tree, then x plays to block, and then it is not hard to see that x wins, as he has two ways of creating a line. If o plays as in the center option, then x must also block, but this path leads to a draw with perfect play on the part of both players. Finally, the rightmost path also leads to a win for x by a similar analysis to the left path. From this diagram, it is clear that o must take the middle path. He will then either draw with perfect play, or perhaps even win if x makes a mistake.

This analysis is indicative of the kind of reasoning accompanying search, although by no means reflects the complexity of this technique (carnival chickens can play a perfect game of tic-tac-toe). A full-fledged treatment of search would include a systematic way of propagating the evaluations of the positions from the leaf nodes to the root node of the tree. In addition, there are many other types of search besides those appropriate for game playing. The interested reader is referred to any introductory text on Artificial Intelligence, which will typically contain an exposition on search near the start of the book.

Position evaluation is crucial to game playing, as it is usually not possible to reason to the end of the game. This is typically done with what are known as heuristics, or rules of thumb. These rules provide a numerical value for each position, and in concert with the proper search techniques, form the backbone of most game-playing systems. System search2 on the CD-ROM provides an evaluation system for the first level of response in the game tree in Figure 9.13(B). Advancing the simulation once shows the system's evaluation for each of the open possibilities for the x (light blue), with the deeper red indicating greater worth. The system suggests that he should play in one of the corners in this position.

ON THE CD

A collection of such evaluations could be used as the basis of a game-playing system. It should be noted, however, that in a complex game like chess or Go, the fact that a move receives a high evaluation is not sufficient evidence to play there. Further search must be done to see if the opponent has a response (and to check the responses to this response, and so on) that renders the seemingly good move inadequate. In other words, game playing and most complex situations are not matters of pattern recognition alone but require pattern recognition to act in the service of generalized search algorithms.

EMERGENT BEHAVIOR

Finally, we treat a topic that is not so much a technique in the arsenal of those modeling intelligence as it is a property that is desired to be present in all methods. Emergent behavior refers to the idea that there is synergy in the system in the sense that the whole is somehow greater than the sum of its parts. This is especially desirable in the case of intelligent systems because we are trying to achieve so many things at once. Behaviors that are almost automatic for humans, for example, such as language comprehension and understanding everyday physics (for example, preventing liquids from spilling and judging trajectories of moving objects) have proven exceedingly difficult to program because of the sheer quantity of these types of events.

These problems and others have given impetus to approaches that are biologically motivated because it is apparent that emergence is a property that exists both at the level of the individual organism and the species as a whole, regardless of the animal in question. Individual organisms invariably possess a number of automatic behaviors that are complex amalgams of the environment, their bodies, and their peripheral and central nervous systems. If this nexus could be captured and understood, then adaptive behavior at the very least could be better modeled, and perhaps intelligence itself could be approached. After all, the reasoning goes that human beings are members of the animal kingdom also.

One property of all living systems is that they learn to adapt. You may recall from the earlier section on neural networks, Rosenblatt's belief that his Perceptron, in a suitably complex environment, might achieve some near-humanlike capabilities. This dream has been somewhat tarnished, not so much by the limitations on the abilities of neural networks to learn, but on the basis of any system's ability to learn. It may be a slight overstatement, but there is a great deal of validity to the notion that in order for a system to learn something well it must already have an almost complete understanding of the concept it is trying to acquire. We learn geologically, with each successive step providing a new layer of knowledge on top of the existing knowledge base. The blank slate approach, in which a completely neutral system could come to be intelligent through mere exposure to the world and the proper learning algorithms, has fallen largely into disfavor.

Still, the idea of emergence of some sort, is an enormously appealing one. That it may be possible to get a large collection of relatively simple parts to emulate complex behaviors remains an appealing avenue of research. We have already seen examples of such collections, known as complex systems, in neural networks and in systems based on cellular automata such as LATTICE. In Chapter 1, we introduced Conway's *Game of Life* and we showed many emergent behaviors including stability, gliders, and repetition. None of these patterns of activity are explicitly coded for; they are all emergent properties of the system in action.

ON THE CD

Systems emerge1, emerge2, and emerge3 on the CD-ROM are similar to the *Game of Life* but extended to state-based automata. Table 9.2 shows some of the emergent behaviors.

Among other things, this table illustrates one characteristic of complex systems, cyclic behavior. This behavior is a function of both the initial conditions and the rules that drive the system. For example, emg-pat4 has a tendency to produce cyclic behavior, as does system emerge2. If the behavior is not cyclic, then the pattern will either expand or die. Another property of complex systems is sensitivity to initial conditions. Try adding extra cells to a pattern or altering the rules for a system slightly. In many cases, these small alterations will lead to radically different behavior.

An additional emergent property of complex systems such as LATTICE is repeating behavior. We distinguish between three types. The first, known as a fixed-point attractor, occurs when a subsequent pattern repeats itself exactly. In the *Game of Life* (system life in the chapter1 examples folder), there are many such examples. The second and somewhat more interesting type is the periodic attractor, in which

TABLE 9.2 Emergent Behaviors in Three Variations on the *Game of Life*

	emerge1	emerge2	emerge3
emg-pat1	exploding diamond	cycle	dies
emg-pat2	explodes then dies	cycle	dies
emg-pat3	expanding and imploding blobs	dies	dies
emg-pat4	cycle	cycle	horizontal movement, then cycle
emg-pat5	irregular gliders	dies	cycle

the patterns cycle through various states regularly; this was also seen in the *Game of Life*. Finally, a less well-known, but from our point view considerably more interesting, type of periodicity is what we may call the "musical attractor." The musical attractor is characterized by nonsimple repetition, in which slowly evolving cycles within cycles may or may not return to the original state. The reason for deeming this sort of behavior musical is that this is just the kind of periodicity we see in human-composed music, and therefore presumably reflects human musical tastes. Too much repetition tires the ear, but too much complexity leads to a lack of unity in the piece. The musical attractor provides just the right blend of repetition and variation to be pleasing. The system below is an example of this effect.

ON THE CD

System boole on the CD-ROM contains the LATTICE composition "Blues for Professor Boole." This composition is based on the blues scale of c-d-e^b-f-g-a-b^b-c. Push the speed slider to the maximum value; this will allow the beat cell to control the tempo completely. You can change the melody by altering the contents of the random machine (this can be done while the simulation is running), or by altering the state tables if you desire a more radical change. In most cases, you will find that the repetition is neither periodic nor random but some musical-like combination in between.

While it is nice to have systems with emergent properties, the challenge with respect to intelligence is to get the right sort of properties to emerge. Intelligence, as has been argued, is not mere complexity—after all, a purely random pattern is highly complex—but a judicious blend of information-generating properties. In certain cases, such as music composition, reasonable-sounding compositions are possible because the system possesses the right kind of periodicity for music. It is unclear whether emergence will be sufficient for other, or even a minority of intelligent behaviors. Nonetheless, emergence remains an intriguing property for those attempting to build intelligent machines.

SUMMARY

We began this chapter with a working definition: that intelligence implies the ability to add information to one's database of knowledge over and above the information directly provided by the environment. Then we explored various mechanisms that conform to this definition. We saw that pattern recognition, or the ability to classify objects in one's environment, is a critical component of any intelligent system. This yields information based on the category the object is placed in. For example, if we recognize a dangerous animal ahead, we know we should be on guard. Then we considered pattern completion, or the addition of stereotypical information to an object which is only partially in view.

Pattern completion is generally thought of as a perceptual process. Higher-level addition of information is usually couched in terms of inference. Inference typically, although not necessarily, is accomplished with Horn clause rules or IF-THEN rules. Such rules use the implication operator and allow the conclusion to be made when all the antecedent conditions are met. Various relaxations of these rules however, are necessary if they are to be useful in expert systems, or programs that make complex decisions. In addition, we discussed how chains of rules are more powerful than a single layer system.

We discussed neural networks, which are not really a single mechanism, but a general representational system for accomplishing a host of tasks. We know that at a minimum, intelligence can be realized with such networks as we have an existence proof in human beings, although it is not yet established that human neural networks work similarly to artificial networks. These networks *are* excellent models of parallel processes such as vision. They can also be used for learning and adaptation. Rather than treat this in detail however, we discussed learning with a general model that was remarkably similar to an earlier topic: minimization on the Karnaugh map.

We concluded with a brief treatment of search, and then looked at the possibility that intelligent behaviors could emerge from the collective behavior of simple elements. The LATTICE system is an exemplary instance of this, and with

LATTICE we were able to model a wide range of digital techniques in this book. The last LATTICE system was a composition in honor of the founder of mathematical logic entitled "Blues for Professor Boole." Really, though, the good professor has nothing to be sad about. He indirectly laid the mathematical foundation for digital computation, and without his work the array of digital devices that have been the subject of this book, and that now permeate our lives, would scarcely be possible.

EXERCISES

LATTICE Exercises

9.1 Create a system in LATTICE, along the lines of PR1 and PR2, to detect the occurrence of the letters T and S. Do this in two ways, first by looking for a proportion of pixels in the whole letter and then in a two-stage process whereby constituent features are first extracted and then PR is performed on those features.

9.2 Create a system along the lines of PC1 that restores as many patterns as possible to the form shown in Figure 9.14. Make sure you turn off the center cell if it is on in addition to turning on the outside edges if they are not fully present.

FIGURE 9.14 The target pattern for Exercise 9.2.

9.3 Augment system PC2 with the ability to restore partially present diagonal lines in addition to the horizontal and vertical line restoration already in the system.

9.4 Show that $(P \Rightarrow Q)$ is equivalent to $(P' + Q)$. Construct a circuit that corresponds to this connective with only NAND gates. Design and implement in LATTICE a simple expert system that decides whether you should hire a prospective employee as a digital designer.

9.5 Augment system net2 so as to activate only the R for the first letter if the entire input word is RIM.

9.6 Create a horizontal line filter and the corresponding horizontal edge filters, add this to the logical expression in net4, and apply it to net4-pat1 and net4-pat2.

9.7 Implement a horizontal generalization rule comparable to that in learn1 and show that it fails in learn1-pat1 because of the negative example. Remove the negative example and show that it works. Then implement a more conservative generalization rule that requires at least four of the cells to be positive examples and no cells to be negative examples in the horizontal bar of eight. Test this new system on learn1-pat2.

9.8 As implied in the chapter, a radical learner may choose to ignore negative examples at times. Implement a system that generalizes in a 2 by 4 horizontal bar if at there are at least four positive examples in the presence of no more than one negative example. Make sure the negative example converts to a positive one by testing your system on learn1-pat3.

9.9 This is a very open-ended problem but may prove interesting to the artistically inclined. The idea is to reimplement learn5 but the goal here is to create the most interesting work of animation art. You can leave out the partitions for the notes and rhythm and just concentrate on the visual aspects of the resulting machine. Once you have found two animations that you like, try combining the genes of both to see what happens. You may also introduce "mutations" by altering a bit or two to see the effect on the animation.

9.10 Create a system along the lines of search2 that rates a move highly any time it blocks blue when he has the possibility of making three in a row on the next move.

9.11 Design a state-based system with one or more partitions representing different instruments. Make sure your result sounds musical, that is, is neither too repetitive nor too complex. If you feel confident of your system, you may submit it to *http://www.universalhedonics.com/lattice*. We will post the best of the submissions (these submissions can be pure animations or animations influenced by mouse-clicks also).

ENDNOTES

1. Those versed in programming may see a close correspondence to this statement and object-oriented programming (OOP), the now dominant strain in computer languages introduced in Chapter 8. OOP, by allowing one to associate data and procedures with object types, allows one to easily make default inferences about objects in the virtual universe that the program is addressing.

2. It is also possible to have them have a lower threshold than other features, which, as Equation (8.2) shows, is essentially the same thing.
3. A famous incident along these lines occurred when the physicist Herman Weyl almost abandoned his gauge theory of gravitation because it was not in accord with all the data. Weyl chose to throw out the data rather than his theory, and his choice was later justified when gauge invariance was incorporated into quantum electrodynamics.
4. Actually, a better catchphrase would be "survival of the fertile," which includes the notion of surviving to mating age and the ability to attract an appropriate mate. These dual factors have long been known and Darwin distinguished between the two in his landmark "Origin of Species."

Appendix The LATTICE System

INTRODUCTION

LATTICE (Logical AuTomaTa Integrated Creation Environment) provides a general framework for the creation of cellular automata (CA)–based systems. CAs are special types of complex systems, that is, systems made up of large numbers of interacting parts. These parts are themselves relatively simple, although the resulting emergent behavior of the complex system as a whole may be difficult to predict and at times difficult to understand even at an approximate level; hence the name complex system. A CA consists of cells whose behavior is completely determined by the contents of cells in the vicinity of the cell in question. CAs have been used, among other things, as models of physical phenomena, as models of mathematical systems, as models of computational systems, and, as in the case of some of the examples in this book, as generators of music and visual art.

Unlike most CAs in which the cells are driven by symmetric expressions, the cells in LATTICE are driven by possibly asymmetric logical and state-based rules. For example, a typical CA rule could be that a cell will die (become inactive) if fewer than two cells in its immediate eight-cell neighborhood are alive. In contrast, cell activity in LATTICE is determined by either logical or state-based rules driven by arbitrary surrounds. In the case of the former, these rules consist of logical expressions, the variables of which correspond to subsets of the surround of the target cell. In the case of the latter, cells move to future states based on the values of variables, which also are triggered by the states of the cells surrounding the target cell.

Another key difference between LATTICE and most CA systems is that LATTICE is partitionable. This means that different subsections of the grid can contain different cell definitions. Within a given partition, cells behave identically, as in traditional systems. Although the partitions may contain different behaviors, cells respond identically whether their neighbors are within their partition or not. The only thing they are responding to is the state of these neighbors at a given time, not how these neighbors evolve. This interfacing between adjoining partitions is necessary for many of the systems in the book. For example, the L-puter, or LATTICE-based computer, is composed of numerous partitions interacting in complex ways to simulate a simple working microprocessor.

State-based systems also differ from traditional systems along a number of dimensions. One such difference is nonlocal detection. The future states of cells can be influenced not only by cells in the local surround, but also by cells in the horizontal or vertical lines extending from the target cell. State-based cells also have the ability to nonlocally copy the contents of distant cells if the appropriate conditions are met. Finally, when cells reach a given state, they may trigger additional actions, either by flashing, playing an arbitrary piece of music, or displaying an image within the cell.

It would be a mistake to claim that these additional capabilities provide LATTICE with greater computational power than ordinary CA systems because it is known that such systems can emulate a Turing machine, the most general computer. Nevertheless, the inclusion of these features makes the construction of CA analogs of digital circuits considerably easier. In addition, and this is the primary motivation behind LATTICE, there is a natural correspondence between the logic-based systems and combinational circuits, on the one hand, and state-based systems and sequential circuits, on the other. In short, mastering LATTICE provides a solid basis for mastering the mathematical foundations behind digital design.

INSTALLATION

System Requirements

Any Windows OS including XP, 2000, NT, ME, and 98 with at least 32 M of main memory and a monitor with at least 1000×600 resolution will suffice. CPU speed is not critical although processor speed will affect the speed at which the animation runs when the animation speed slider is set to its fastest value (at this setting, there is no delay between successive cycles). The Java Runtime Environment, which can be found at *http://www.java.sun.com/getjava* must also be installed. On most new machines it will not need to be installed separately, however.

Installation Procedure

Click on the LATTICE install icon on the CD-ROM and follow the directions. Once LATTICE has been installed, if you receive a message to the effect that the

Java Runtime Environment is not present, or if you receive a message such as "Can't launch: No main class found," you must also install Java on your machine.

General Operation

This section describes the features of the program that are shared between the two types of systems: truth and state, in LATTICE.

Program Layout

LATTICE is divided into two panes. The logical or state definition is on the left, and the cells themselves are on the right. The panes are not resizable, but clicking on the left caret in the divider between the panes will expand the right pane (the animation pane), and clicking on the right caret will expand the left pane (the definition pane). The program can then be restored to its original state by clicking on the right and left carets respectively. Beneath the definition pane is a button panel, the contents of which depend on the type of system. Beneath the animation pane are buttons and controls that are the same for both types of systems. In the upper left is the File menu, which is now described, after which the animation options and the effect of the mouse buttons are detailed.

File Menu Options

New Truth System This creates a new nameless truth system. A name will be requested for the system when it is saved.

New State System This creates a new nameless truth system. A name will be requested for the system when it is saved.

Open System This opens a dialog box in which a system can be opened. Only LATTICE-generated system files can be opened.

Open Pattern This opens a dialog box in which a pattern (a set of cell values in the animation pane) can be opened. Only LATTICE-generated pattern files can be opened.

Save System This saves the current system. If the system is nameless, a name will be requested in a dialog box.

Save System As This opens a dialog box and saves the current system in the requested folder with the indicated name.

Save Pattern As This opens a dialog box and saves the current pattern in the requested folder with the indicated name.

Exit Exit LATTICE.

Animation Options

Cycle   Shows the current animation cycle. This is not editable.

Animation Buttons

- Stop the animation.
- Single step through the animation.
- Step through animation until stop is pressed.
- Reset the animation to start.

Speed Slider   Controls the animation speed. At maximum setting, the animation will proceed as fast as the processor allows.

Granularity Box   Controls the size of the cells. Finer grains will result in slower animations because more cells are present.

Machine Definition Icon   Clicking on this icon brings the program into machine definition mode. In this mode you may define a new machine (cell partition) by dragging from the top left cell of the machine to the bottom right. Note that there is always at least one machine in any system, the background machine labeled "bgrnd." The program will not allow you to create a machine that overlaps with an existing machine (nor will it allow you to start the machine in the top-left cell; this is reserved for the "bgrnd" machine). Once dragging is complete you will be asked to name the machine. This name will appear in the top-left cell. To change the name of an existing machine, click on the machine definition icon, and then click anywhere on the machine. To delete a machine, click on the machine definition icon, then click anywhere on the machine, and then click on "del." Additional options applicable to state systems only are described in that section (under state action, sound file).

Mouse Buttons

In the default mode (not machine definition mode), the mouse buttons have the following effects on the cells in the animation pane:

Left Button   Advance selected cell one color. This will also select a machine if the selected cell is not in the currently highlighted machine. The logical or state definition for the selected machine will appear in the definition pane.

Right Button   Decrement selected cell one color (because logic-based machines only have two states, this will be equivalent to using the left button). This will also select a machine if the selected cell is not in the currently highlighted machine.

Center Button Select the machine corresponding to the selected cell without changing the cell color.

TRUTH SYSTEMS

Truth systems are driven by expressions, one for each machine in the system. These expressions appear in the definition pane on the left. In addition to influencing cell behavior, this pane is effectively a self-contained truth table program. As the expression is modified, the 0s and 1s in each column are automatically adjusted to reflect the correct value for this row and column of the truth table. There are four types of columns in the truth table:

Variables These appear in sequential order on the left of the table and are headed by "var." The truth values appear in the normal binary counting order. Truth values are not editable but variable names are; double-click on a variable to edit it and hit return when the editing is complete. All other columns after these initial columns are part of the truth expression.

Variables in the expression As in the previous category, these variables are headed by "var." However, these variables always contain one or more left or right parentheses, which are generated automatically by the program. The binary digits below the variable reproduce the values from the previous category. Variables may be changed, complemented, or expanded in the manner to be described below.

Operators in the expression Operators are headed by "op." Their truth values beneath the operator are computed as a function of the operator and the other elements in the subexpression containing this operator. Operators may be changed, complemented, or used to collapse subexpressions in the manner to be described below.

The special "F" operator This function operator behaves identically to the other operators except that this column reflects the truth values of the expression as whole. If the expression were laid out as a tree, this operator would be the root of the tree.

Variable Drop-Down Box

Clicking on any variable in the expression will create a drop-down box with three options:

268 Digital Design: From Gates to Intelligent Machines

1. A list of the currently used variables
   Select one of these to change the variable.
2. ~ This complements the current variable. This will flip the truth values for this variable and will affect other values in the expression accordingly.
3. ≡ Expand
   This will expand the current variable to a sum of the variable with itself. For example, if the variable is P3, it will be replaced with (P3 + P3). This subexpression can then be changed as desired.

Operator Drop-Down Box

Clicking on any operator in the expression will create a drop-down box with nine options:

1. +
   Select this to choose the sum (OR) operator.
2. ∗
   Select this to choose the product (AND) operator.
3. ?
   Select this to choose the equivalence operator.
4. NAND
   Select this to choose the NAND (not AND) operator.
5. NOR
   Select this to choose the NOR (not OR) operator.
6. XOR
   Select this to choose the XOR (exclusive OR) operator.
7. ~ This complements the operator. Thus, the truth values in the current column are flipped. Note that the complement appears as a "'", or complement, in the appropriate place in the expression. For example, suppose the original expression was (P1 + (P2 ∗ P3)). Complementing the "∗" will result in the new expression (P1 + (P2 ∗ P3)'). Complementing the "+" will then result in the final expression (P1 + (P2 ∗ P3)')'.
8. ⇐⇐ (collapse left)
   This will collapse the subexpression corresponding to the current operator to the left. For example, suppose the expression is ((P1 + P2) ∗ (P3 + P4)). Selecting this option on the "∗" will transform this expression to (P1 + P2).
9. ⇒⇒ (collapse right)
   This will collapse the subexpression corresponding to the current operator to the right. For example, suppose the expression is ((P1 + P2) ∗ (P3 + P4)). Selecting this option on the "∗" will transform this expression to (P3 + P4).

Truth Table Options (Bottom of Screen)

1. **Machine name**
   The name of the currently selected machine is displayed on the left. This is not directly editable but will change as different machines are selected in the animation pane.
2. **Add var**
   Clicking on this button will add a variable to the left of the truth table. Truth values are adjusted accordingly.
3. **Add clause**
   Clicking on this button will add the new clause "+ P1" to the existing expression. For example, suppose that the existing expression is $((P1 * P2) + P3)$. The new expression will be $(((P1 * P2) + P3) + P1)$.
4. **Color buttons**
   Clicking these will bring up a color dialog box that will allow the inactive and active colors for the cells to be set.

Variable Settings Dialog Box

Figure A.1 shows the variable setting dialog box that appears when any "var" is clicked in the top row of the table. This dialog box allows one to define the conditions under which each variable becomes active (logic 1). The options for this box are as follows:

1. **Variable**
   This allows the selection of the desired variable from the current list of variables.
2. **Triggers**
   Clicking on these cells selects the possible triggers in the surround of the current cell (the marked center cell), which are needed to be active to activate this cell (unclicked cells with the off color are ignored). Any or all of the cells in the indicated neighborhood of the center cell may be chosen, and the center cell itself may also be chosen. If it is necessary that this cell is on to keep the current cell active on the next step, the "center necessary" box should be checked.
3. **Bounds**
   This controls the number of the trigger cells that must be active. For example, suppose five trigger cells have been chosen in the previous step. Choosing the bounds "from 2" and "to 4" means that when greater than or equal to two and less than or equal to four of these five cells are active, the current cell will become active on the next time step. If the bounds are left in the default state, "ALL" and "ALL." then all five must be active.

270   Digital Design: From Gates to Intelligent Machines

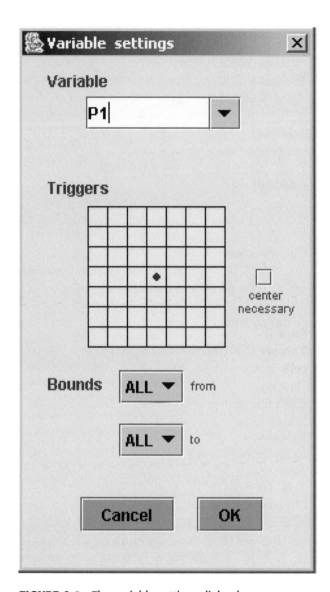

FIGURE A.1   The variable settings dialog box.

STATE SYSTEMS

State systems differ from truth systems in that cells transition from their current state to any one of multiple states, each with a color designation. There is a unique transition for each of the truth values of the variables. For example, if there are three variables, there will be $2^3 = 8$ transitions for each state. In addition, each state may be attached to an action, as described below. In the table, each row corresponds to a unique state. Columns are organized in the table as follows:

Column 1 (State Color)

This defines the color for the state. The default colors may be altered by clicking on this column, which brings up a color dialog box.

Column 2 (State Action)

Clicking on this column brings up the action dialog box shown in Figure A.2. This allows one to define the action that will be taken if this state is reached. Options are as follows:

1. **Fast flash**
   This will cause all cells in this machine to flash quickly upon reaching this state.
2. **Slow flash**
   This option, which is mutually exclusive with the previous one, causes all cells in this machine to flash slowly upon reaching this state. This may slow down the animation considerably in the case where many such cells reach this state at once.
3. **Sound file**
   Selecting a sound file by browsing will cause this sound file to be played when this state is reached. Sounds play in a separate thread and do not have a significant effect on animation speed. In the case where multiple cells in the same machine reach the same state at once, only one instance of the sound will be played (if multiple cells in the machine reach different states that trigger sounds, then all of the sounds will be played at once). Formats currently supported are .wav and .au files.

   Machine settings also influence the production of sounds. In the machine definition box, the clip sounds check box will suspend the sound at the start of the next animation step. Otherwise, the sound will continue (unless the new state is connected to a different sound). Leave this box unchecked if you want a single loop to play continuously throughout the animation. The beat box check box appears for single cell machines. If this box is checked, then the beat for the entire system will be driven by the length of the sound being played by this machine. For example, you can use silent beats to control the rhythm of the produced music (see the music system in Chapter 1 for an example of this use). Only one single cell

machine is allowed to be the beat box; checking this for one machine will turn off this option for all other machines.

4. **Picture file**

   Selecting a picture file by browsing will cause this image to be displayed in the cell when this state is reached instead of the color corresponding to this state. The image will automatically be adjusted to fit in this square (this means that if the original image is not square, it will be stretched in the horizontal or vertical direction to fit the cell). Formats currently supported are .jpg, .gif, and .png.

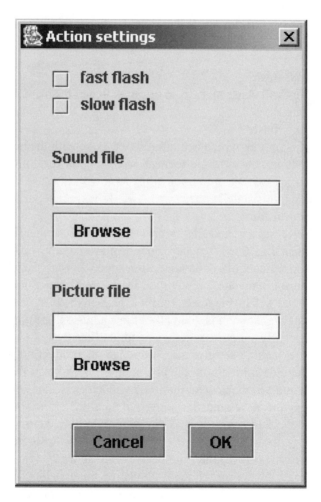

FIGURE A.2 The action settings dialog box.

Column 3 (State Name)

This column contains the names of the states. These names may be edited by double-clicking on the cell. The altered name will be propagated throughout the table.

Columns 4 Through the End of the Table (State Transitions)

Cells in these columns represent transitions. The drop-down box in each cell contains a list of all states that may be selected from. In addition, if there are one or more machines in the system with the same dimensions as the current machine, there will be the option to copy from these machines. These are indicated by "⇐" and the name of the machine to copy from. When this option is selected, the current cell will receive the state values of the corresponding cells in the chosen machine when this transition is effected instead of moving to a fixed set of states.

State Variable Settings

Clicking on the top row of the variable columns (labeled with the variables $In...I0$) brings up the state variable settings dialog box. This determines when the chosen variables become logic 1, which in turn affects the transitions via the transition table. The options for this box are as follows:

1. **Variable**
   This allows the selection of the desired variable from the current list of variables.
2. **Triggers**
   This works analogously to the corresponding dialog box for truth systems except that cells may be set to any of the current state colors. The default setting of each cell is neutral (light gray) meaning that the state of this cell will be ignored. Left-clicking moves forward through the list of states and right-clicking moves backward. As before, the center necessary box may be checked to indicate that the current cell must be in the indicated state.
3. **Bounds**
   Bounds also work analogously to the truth system dialog box. In this case, the bounds refer to the number of nonneutral (light gray) cells that must be active.
4. **Line triggers**
   Line triggers are mutually exclusive to the above triggers. Setting a non-neutral color on any of these will look for the color anywhere in the animation pane. For example, setting the right line trigger to be in a given

state will activate the given variable if *any* of the cells to the right of the cell in question are in this state. These triggers are disjunctive, i.e., any one or a combination of the active triggers is sufficient to set the variable to logic 1.

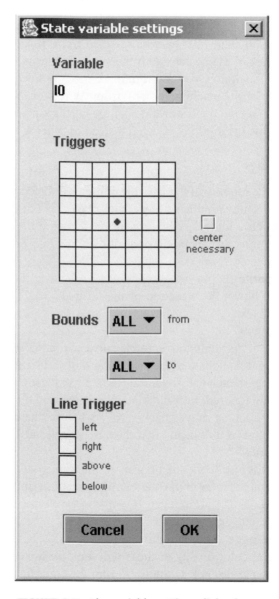

FIGURE A.3 The variable settings dialog box.

State Table Options (Bottom of Screen)

1. **Machine name**
   The name of the currently selected machine is displayed on the left. This is not directly editable but will change as different machines are selected in the animation pane.
2. **Add state**
   This will add a new state to the system. States are shared by all machines in the system so this will implicitly update every machine. A new color will automatically be chosen for the state although this color can be changed by clicking on the first column.
3. **Add var**
   This will add a new variable to this machine. Because every possible truth value of all variables is shown in the transition table, this will double the number of columns after the third column. Variables are automatically assigned the names I0, I1, etc.

TRICKS OF THE TRADE

The best way to learn how to use LATTICE is to play with the examples in the book and to construct systems of your own. However, there a few tricks that may help in system design.

1. **Constant cells**
   If you wish all cells in a given machine in a truth system to remain constant, simply use the default expression (P1 + P2) and set the trigger for P1 to be the center cell (P2 is ignored). This means that a cell will only be active if it was previously active, effectively causing it to remain in the same state. In a state system, constancy can be provided by using a single variable (the default state) and selecting the same states in the first variable column (I0 = 0) as the state for this row. This means that the state will progress to itself when I0 = 0, which it will do by default because I0 is not defined.
2. **Moving cells**
   To get a cell to move in a given direction in a truth system, set the trigger for P1 to be the opposite of the intended direction. For example, to get all active cells to move right, set the trigger to be a single cell to the left of the center cell. This will cause the next state of any cell to the right of an active cell to become active. To create a line of active cells to spread out from an active cell, use the default expression (P1 + P2), keep P1 as before, and

set P2 as in the constant case described above. To get cells moving in a given direction in a state system, create a new variable for each state that you wish to move, and set the corresponding transition in the top row of the table to be these states. For example, to make any cell in state B move to the right, and any cell in state C move diagonally to the up and left, set the trigger for I0 to be a single cell to the left of center to state B, and the trigger for I1 to be a single cell to the bottom right of center to state C. Then set the top row of the state table to be ABCA, in that order.

3. **Traditional CA rules**

    Traditional CA rules are symmetrical and are phrased in terms of creating and maintaining life (the active state). To create these in LATTICE, use symmetrical triggers, and exclude the center if the rule is about creating life, and include it as necessary if the rule is about maintaining life. For example, examine the triggers for P1 and P2 in the life system. P1 effectively states that a cell will come to life if any three of the cells in the immediate surround are active. P2 effectively states that a cell will remain alive if it is already on and two or three cells in the immediate surround are also active. Note that the LATTICE rule includes the center cell in the count and thus the bounds are between three and four.

4. **Line triggers**

    Line triggers are useful when you wish to enable an operation that applies to every cell in a machine and the machine is bigger than the normal trigger bounds. For example, examine the definition of I1 for any of the memory units M0 to M15 in the Lputer system (see Chapter 8). It would be awkward to have a trigger that causes copying from the bus because it would need to operate on every cell in this eight-cell machine. The line trigger allows every cell in this machine to work in the same way.

5. **Copy from**

    Copy from is useful in any nonlocal transfer of contents. For example, it is used extensively in the L-puter system as a means to transfer data from the bus to memory, and to copy the control signals from one part of the system to another. Once copied, these contents may then be used by local triggers.

SYSTEM SUBMISSION

If you have created a LATTICE system that makes particularly interesting music and/or contains good visual effects, and you wish to share this with the world, please visit *http://www.universalhedonics.com/lattice.html* for submission instructions. We will post the best of the submissions so that they can be run online along with your name and the title of your composition. Universal Hedonics reserves the right to refuse posting for any reason.

Index

A

accumlator. See L-puter
address register, 190. See also L-puter
ALU, 181, 189–91, 202, 213, 214, 217, 218. See also L-puter
analog versus digital, 1–2
application-specific integrated circuit, 107
Arithmetic Logic Unit. See ALU
artificial intelligence, 229–30, 248. See also emergent behavior; learning; neural networks; search
ASCII, 11, 20–21
ASIC. See application-specific integrated circuit
Asimov, Isaac, 118

B

backprop, 247–48
bases. See also binary number systems
    conversions between, 8–10 (see also BCD; Gray coding; parity problem)
    decimal equivalents for common nondecimal, 7
    defined, 6–7
    divisors, 8
    ideal, for computing, 70
BCD, 18–19
binary adders
    carry–look-ahead adder, 86–88
    full adder, 82–84
    half adder, 84
    ripple-carry adder, 85–86
    two's complement addition and subtraction, 88–90
Binary Coded Decimal. See BCD
binary number systems. See also ASCII; Unicode
    arithmetic operations, 12–18 (see also binary adders)
    full count, 12
    negative numbers, 15–18
    powers of two, computing significance of, 11–12
    representing logical processes, 11
    signed magnitude, 15
    two's complement, 16–18 (see also binary adders)
Boole, George, 26
Boolean algebra. See DeMorgan's law; logical operators; maxterm; minterm; POS; prime implicant (PI); simplification of algebraic functions; SOP; tautology; truth table
bottom-up processing, 243
bus, 182–83, 210
ByteCode, 228 n. 1

C

C, 202
C++, 226
carry-look-ahead adder. See binary adders
CAs, 2–3, 257, 263
cascading
    decoders, 94–95
    memory, 189
cellular automata. See CAs
central processing unit. See CPU
chips. See ICs
Church-Turing hypothesis, 180
circuit delays, 72–74, 88
circuit design, technological constraints on, 67–69. See also cascading; circuit delays; fan-in; fan-out; gate delays; noise; voltage
circuits. See also combinational circuits; finite state machine; gates, elementary; iterative circuits; latches; sequential circuits
    corresponding logical functions, 59–61
    corresponding truth tables, 60–61
    economy of design, 34, 66–67, 74–77, 84, 85, 86, 92, 140

279

gate inversion, 64–65
realizing from functions, 61–63
realizing SOP and POS functions, alternative representation of, 64–65
realizing through minterms and maxterms, 63–64
CISC, 225–26
clock functions, 218. See also flip-flops; latches; registers; sequential circuits
clock-triggered D latch. See latches
CMOS, 19–20, 68–69, 71, 73
codes. See also BCD; Gray coding; parity problem
defined, 18
combinational circuits, 56, 59–61, 117, 119. See also binary adders; circuits; decoders; demultiplexers; encoders; gates, elementary; multiplexers; PLDs
Complementary Metal Oxide Semiconductor. See CMOS
complex instruction set computer. See CISC
computer
flexibility, 179–81
memory, 183–89
organization, 182–83
programming, 180–81
control unit, 189–90, 204–5, 224
data paths, 210
fetch-decode-execute cycle, 204, 205 (see also L-puter)
finite state machine, 207
Conway, John, 3
CPU, internal organization of, 189–91. See also interrupts; microsequencing

D

data lines. See bus
data register, 190. See also L-puter
decoders. See also cascading; demultiplexers; encoders
applications, 95–97, 187–88
binary decoders, 91–95
decoding, 90, 190
DeMorgan's law, 32, 41, 64
demultiplexers, 100, 105–6

design economy, 34, 66–67, 74–77, 84, 85, 86, 92, 140
deterministic finite state machine. See DFSM
DFSM, 140–41, 177n
digital circuits
digital design, defined, 1–2
Dozenal Society, 8
dynamic RAM (DRAM), 186

E

edge-triggered flip flops. See flip-flops
emergent behavior, 256–59
enable line, 91
encoders, 97–100. See also decoders
encoding, 90
ENIAC, 68
equivalence, 31–32
erasable PROM, 107
expert systems, 236–40

F

fan-in, 70–71, 94
fan-out, 70–72
field-programmable gate array, 107
finite state diagram, 150–51
finite state machine, 140–44, 149, 157, 207. See also control unit; L-puter
flip-flops, 125. See also parity problem; registers; sequential circuits
D flip-flop, 125–27, 170–71, 173, 174
J-K flip-flop, 127–29, 171–74
FPGA. See field-programmable gate array
full adder. See binary adders

G

Game of Life (Conway), 3–4, 257–58
game-playing systems, 255–56
gate delays, 72–74
gates
elementary, 56–59 (See also circuits)
implementation of, 74–77
inverted, 64–66, 74
general register. See L-puter
generator, 86

Gray coding, 19–20

H
Horn clause rules, 237–38

I
ICs, 68
incrementing hardware, 200
inference, 236–39
information storage, 11–12, 88, 180, 181, 183–89. See also registers
instruction register. See L-puter
instruction sets, 197, 201–4, 225–26
integrated circuits. See ICs
intelligence, nature of, 229–31
intelligent systems, 170. See also emergent behavior; expert systems; learning; neural networks; search
interrupts, 224–25
inverters, 42, 56, 60
I/O operations, 191–93
iterative circuits, 85, 86
iterative loops, 203

J
Java, 202, 226, 228 n. 1

K
Karnaugh maps, 42–50, 152, 158–59, 161, 209, 249

L
latches
    D latch, 123–24, 125
    SR latch, 119–22
    transparent latch, 124
LATTICE, 4, 13, 28, 29–30, 62, 108, 141–42, 144, 170, 246. See also L-puter
    CA rules, traditional, 276
    cells, manipulating, 275–76
    characteristics, 2–3, 263–64
    copying within, 276
    installation and general features, 264–67
    line triggers, 276
    state systems, 270–75

    truth systems, 267–70
learning
    genetic algorithms, 252–53
    supervised, 249–50
    unsupervised, 250–52
logical automata, 2–3
Logical AuTomaTa Integrated Creation Environment. See LATTICE
logical gates. See gates, elementary
logical operators, 26–29, 31, 33–34, 42. See also circuits; gates, elementary
logic technologies, 68–69
L-puter, 197, 218–23, 224. See also control unit
    ALU, 214–17
    control unit
        data paths, 210–14
        DFSM, 207–10
        fetch-decode-execute cycle, 205–7
    instruction set, 201–4
    register set, 198–201

M
maintain state, 119–21
maxterm, 37–42. See also circuits
mazes, 162–70
Mealy machine, 144–45, 147, 151, 165
memory. See information storage
memory, volatility of, 195 n. 1
memory hierarchies, 130
memory-mapped IO, 191–92, 218, 222
memory state. See maintain state
metal-oxide semiconductor field-effect transistor. See MOSFET
microcode, 224
micro-PC, 224
microsequencing, 224
minimization of functions. See Karnaugh maps
minimized circuits, 66–67, 177n
Minsky, Marvin, 247
minterm, 37–42. See also circuits; Karnaugh maps
Modus Ponens, 237
Moore machine, 144–45, 147, 151, 152, 153
Moore's law, 226
MOSFET, 68, 75

multiplexers, 100–105

N

NAND gates, 42, 56, 57, 58, 60, 67, 74–77, 92–93, 95, 124, 174
Necker cube, 241–43
negating positives in binary, 16–18, 89
neural networks, 235, 240–48. See also learning; search
noise, 69–70
NOR gates, 76–77
NOT gates. See inverters
number systems. See also bases; binary number systems
    defined, 4–5
    positional, 5–6
numerals, 4–5

O

object-oriented programming (OOP), 261 n. 1
odd parity. See parity problem
overflow, 17–18, 90

P

PAL. See PLDs
Papert, Seymour, 247
parity problem, 19, 20, 140–42, 153–57
pattern completion (PC), 231, 235–36. See also neural networks
pattern recognition (PR), 231, 232–35. See also neural networks
Perceptron, 247, 257
peripherals, computer, 182
pipelining, 225–26
PLA. See PLDs
PLDs. See also field-programmable gate array
    programmable array logic (PAL), 108–9
    programmable logic array (PLA), 109–12
    programmable read only memory (PROM), 107–8, 184
POS, 37–42. See also circuits; Karnaugh maps
predicate calculus, 26
prime implicant (PI), 44–48, 249–50
priority encoder, 99

product of sum. See POS
program counter, 190. See also L-puter
programmable logic devices. See PLDs
programming languages, 180–81, 201–2, 226
programs, terminating, 222
PROM. See PLDs
propagator, 86–87

R

radix point, 6
RAM, 130, 183–86, 192, 199, 202–2–3, 225
recursion relation, 87
reduced instruction set computer. See RISC
registers
    CPU, 190–91
    parallel-load register, 130–32
    serial-in parallel-out register, 133
    shift register, 132–34
register set, 198–201
ripple-carry adder. See binary adders
RISC, 224–25
Rosenblatt, Frank, 247, 257
rule-based systems, 237, 239, 240

S

search, 254–56
sequential circuits. See also finite state machine; latches; registers
    analysis of, 145–50
    synthesis of, 150–53
        flip-flops, 170–74
        mazes, 162–70
        parity problem, 153–57
        sequence recognition, 157–62
set-reset latch. See latches
simplification of algebraic functions
    Karnaugh maps, 42–51
    logical substitution, 35–37
smart devices
SOP, 37–42. See also circuits; Karnaugh maps; PLDs
state diagrams, 143
static RAM (SRAM), 186
storage. See information storage

sum of product. See SOP
switching algebra, 26
syllogism, 25–26

T
tautology, 31
TDM. See time division multiplexing
time division multiplexing, 104
top-down processing, 243
transistor-transistor logic. See TTL
transparent latch. See latches
tree graph, 61–63
tristate buffer, 183–86
truth table, 27, 29–30. See also circuits

TTL, 68–69
two's complement addition and subtraction. See binary adders

U
Unicode, 11, 21–22

V
voltage, 69–70

W
Weyl, Herman, 262 n. 3